Marketing for Sustainable Development

SCIENCES

Agronomy and Food Science, Field Directors –
Jack Legrand and Gilles Trystram

Food Chain Management, Subject Head – Jean-Marc Ferrandi

Marketing for Sustainable Development

Rethinking Consumption Models

Coordinated by
Sihem Dekhili

iSTE

WILEY

First published 2021 in Great Britain and the United States by ISTE Ltd and John Wiley & Sons, Inc.

ISTE Ltd
27-37 St George's Road
London SW19 4EU
UK

www.iste.co.uk

John Wiley & Sons, Inc.
111 River Street
Hoboken, NJ 07030
USA

www.wiley.com

Library of Congress Control Number: 2021940271

British Library Cataloguing-in-Publication Data
A CIP record for this book is available from the British Library
ISBN 978-1-78945-036-1

ERC code:
LS9 Applied Life Sciences, Biotechnology, and Molecular and Biosystems Engineering
LS9_5 Food sciences (including food technology, food safety, nutrition)
SH2 Institutions, Values, Environment and Space
SH2_6 Sustainability sciences, environment and resources
SH1 Individuals, Markets and Organisations
SH1_10 Management; marketing; organisational behaviour; operations management

Contents

Foreword

John THØGERSEN
Department of Management,
Aarhus University, Denmark

Since the Industrial Revolution, humanity has been extremely successful in combating diseases, producing a sufficient supply of food and other necessities and adapting its environments to its needs. However, this success has a downside. The boom in human production and consumption has led to planetary boundaries for safe operating spaces being crossed in a range of areas, including climate change, biosphere integrity, biogeochemical flows and land-system change. Humanity is now so plentiful and powerful that our activities impact basic planetary functions. This development is so radical that scientists speak about a new geological epoch: the Anthropocene. Therefore, humans, for their own sake, must become stewards of the planet and get us back into a safe operating space, while maintaining acceptable ways of life, as expressed, for example, in the UN's Sustainable Development Goals. This is a major task and it is the government's responsibility to organize and regulate it. However, governments will not be able to prevail without the support and active engagement of companies and the civil society as both consumers and citizens.

Engaged citizens are a valuable source of knowledge and ideas for new norms and regulations that are adapted to the national and local context. An informed citizenry is also a prerequisite for achieving the necessary acceptance and support for new regulations. In addition, changes in consumer behavior are a prerequisite for many transformations, including the move from fossil to renewable energy, from a linear to a circular economy, and to a more plant-based diet in industrialized countries.

The scientific understanding of global challenges and technical solutions has increased tremendously in recent decades, but knowledge and understanding of "the demand side" is lagging behind. We need more knowledge and understanding of citizen-consumers' concerns, limitations, goals and wants with regards to new and sustainable products, services and wider solutions, as part of a wide range of sustainability transformations. We also need more knowledge of how to gain the acceptance, support and engagement of the public, as citizens and consumers with diverse values, needs, wants, resources and contexts.

For example, in developed countries, private households are responsible for about a third of wasted food and about two thirds of plastic waste. Packaging waste, most of which is discarded by households, creates significant problems in nature. Therefore, consumers need to be actively engaged in closing the loops for materials that pass through private households in huge volumes for the radical and urgently-needed transformation from a linear to a circular economic model to succeed. Marketing, as a discipline and practice, has accumulated experiences and insights and developed effective tools to strengthen consumer acceptance of recycled products or products made using recycled materials, and to increase the amount and quality of waste materials that are reused or recycled from households.

This book coordinated by Sihem Dekhili is a welcome contribution to advancing our knowledge and understanding of the role of the demand side for sustainable development and especially about how to mobilize the tools, techniques and insights of marketing for sustainable development. It offers a range of fresh perspectives on sustainability transformations in the modern digital era, drawing on the creativity and skills of a broad group of researchers. Like marketing in general, its main focus is on individuals as consumers and on creating value for all parties in an exchange as a means to achieve organizational and societal goals. This customer-centric perspective of marketing may be the most important contribution to speeding up sustainability transformations. However, marketing is not limited to commercial exchanges and viewing people as individual consumers. Marketing has proven to be an effective means for attaining massive changes in behaviors and lifestyles, including making citizens aware of the need for sustainability transformations in order to accept the required regulations. It is important to ensure consumers are well-informed and understand and trust sustainable products, services and solutions, both to convince them that it is worth their effort and help them to adopt more sustainable goods and practices. Especially, marketing has refined effective tools to help consumers make sustainable choices in supermarkets, including credible sustainability labeling.

Mobilizing consumers and engaging them in sustainability transformations requires a deep understanding of their diversity, and the ambiguity and conflicts related to their goals. Some consumers resist the conventional market system and experiment with various forms of simpler, sufficiency-oriented lifestyles. Others make an effort to choose environmentally-friendly products and services. Still others are environmentally concerned, but feel unable to do anything because they feel that they lack credible environmental information or believe that the tradeoffs are insurmountable. It is therefore important to differentiate between people with different needs, wants, and abilities and to adapt regulation, education, communication and solutions accordingly. More than any other discipline, marketing has developed insights and effective tools for the segmentation and targeting of consumers with different needs, wants, and abilities.

This book is a much-needed contribution to the understanding of the demand side in sustainability transformations and especially of marketing as a force for change towards sustainable development goals. It combines a solid foundation in the accumulated insights of marketing with an appreciation of the specific challenges and opportunities of the current age, including digitalization, mobile applications, machine-to-machine communication and the Internet of Things. These new technologies are rapidly changing our lives and when they are used well, they offer new opportunities for supporting responsible consumer behavior and sustainability transformations. This makes this book a useful resource for marketing scholars and practitioners alike; indeed for everyone who is engaged in the sustainable transformation of society, in companies, politics, NGOs and the civil society.

Acknowledgments

Sihem DEKHILI
CNRS – BETA, University of Strasbourg, France

The journey from an idea born several years ago to the production of this book has been an extremely exciting adventure!

First of all, my warmest thanks go to the 41 authors of this book who have shared my enthusiasm for the topic of responsible marketing and its role in strengthening the sustainable development movement. The exchange of ideas and discussions has been a source of great richness.

All of the authors have brought their expertise to the reflections within the framework of a collective work that has been undertaken in a spirit of attentive listening and conviviality. This kind of project makes the job of an academic even more stimulating.

Huge thanks go to John Thøgersen for the Foreword, as well as for his availability and great kindness. He is a renowned researcher, whose activities and publications in the field of sustainable consumption are numerous.

I would also like to extend particular thanks to Jean-Marc Ferrandi and Patrick Gabriel for their thoughtful advice.

Lastly, the aim of proposing a work anchored in action would have been impossible without the numerous practitioners who offered their viewpoints and enriched the analyses of the researchers. If only they could all be thanked here for their precious contribution!

I hope the readers take as much pleasure from the reading of this book as its authors did from creating it!

Introduction

Sihem DEKHILI
CNRS – BETA, University of Strasbourg, France

For a number of years, sustainable development has been an omnipresent issue in both media discussions and in political, economic and academic debates. It is pushing a real challenge into the spotlight: the balance between the economic, environmental and social components, with the aim of satisfying the needs of the current generations without compromising the ability of future generations to meet their own needs (Brundtland 1987). Sustainable development, looking to the long-term, promotes altruistic values that are beneficial not only to the protection of the planet, but also to social justice and the well-being of others. Indeed, as stated by Gabriel (2003), the wealth created by companies can have an environmental and social cost, and those that benefit from this wealth are not necessarily those that bear the cost. Sustainable development also refers to the idea of controlled production and mindful consumption, with personal pleasure pushed to the background.

Various efforts have been made to take these considerations into account. Most countries have implemented regulations supporting sustainability. Companies have become involved and some have even put sustainability at the heart of their business models. We have thereby seen the development of a new offer on the market, with products using fewer resources and polluting components, and generating less waste (Auger and Devinney 2007; Yannou-Le Bris *et al.* 2019).

At the same time, there has been an increase in environmental and social awareness among consumers. A large number of surveys support the idea that individuals, especially in Western societies, are concerned about ecological crises. This is the case in a survey conducted by Ifop in late 2019, which estimated that

Marketing for Sustainable Development,
coordinated by Sihem DEKHILI.

86% of French people were aware of this issue. Groups of "consum'actors", looking for meaning in life, have emerged and reveal a significant expectation of societal change. As a result, they are turning to an alternative mode of consumption and practices, such as donation and sharing that aim, in particular, to extend the lifespan of products.

The efforts made and the prevailing discussions around sustainable development tend to suggest that we are experiencing a "green era" (Davies *et al.* 2012). However, the reality is quite different! The sustainable criterion is far from being a decisive factor in consumer purchasing decisions, even for the individuals with the greatest level of ecological awareness. Sustainable consumption, primarily the reserve of those with the highest levels of income and education, is for many merely a "surface engagement", as the new consumption pattern that they embrace is being superimposed over entrenched consumerist habits (rather than replacing them) (Daumas 2020). Within this context, the gap between values and declarations, on the one hand, and behaviors, on the other, has been widely documented in the literature, often by using the term "green gap". Despite the efforts made, sustainable development remains the preserve of a niche market. The strategies of companies in this field sometimes turn out to be insufficient. Indeed, their desire to make their image greener can result in opportunist behaviors and greenwashing. Consumers perceive environmental communications as ambiguous and overblown and express little trust in the brands that spread them. Some think that the ecolabeling procedure is based on an incomplete approach that only takes into account a limited number of criteria, and that the communication on ecolabels remains insufficient (Thøgersen *et al.* 2010; Dekhili and Achabou 2015). At the distribution level, super/hypermarkets suffer from a lack of legitimacy in selling ecological products and the price policies are not always considered fair for consumers (Dekhili *et al.* 2017). A cultural barrier can be added to this; responsible practices such as the example of the *doggy bag* can be slowed down by the social norms that dominate a country (Achabou *et al.* 2018).

These obstacles to the spread of responsible behaviors and goods put the importance of the sustainable development movement into perspective. At the same time, they offer immense avenues for exploring solutions to reconsider consumption patterns and develop the green market. This leads us to the key question guiding this book: how can marketing contribute to strengthening sustainable consumption?

Marketing is undoubtedly the area most suited to market development. However, its objectives can be perceived as opposite to those of sustainable development (Kotler 2011). In any case, the compatibility between the two fields has been widely questioned. Marketing aims to sell in a profitable manner. It is seen as a field based on the short term and on the response to selfish motivations. Marketing is also accused of having encouraged overconsumption and waste in shifting the focus from the satisfaction of real consumer needs to a response to short-lived individual desires (Brownlie 2006).

In this respect, the “power” of marketing and its key role in the evolution of modes of consumption should be recalled. Marketing choices can have an influence on individual health through the products promoted and the social groups targeted (as with advertisements for food products high in saturated fats during programs aimed at children). Moreover, marketing affects the representation of individuals and their lifestyles (for example, the demeaning image of women in advertising).

Because marketing enjoys a certain “power” and in view of the significant effects it can have on individuals, particularly the most vulnerable, a number of business practices have been condemned following denunciations from NGOs, scientists and consumer groups. As a result, advertisements that are considered problematic have been withdrawn (as in case of exaggerated claims made by Nivea about its beauty creams) and brands have been boycotted because of sales practices that are considered irresponsible (such as Nestlé in view of its marketing practices for powdered milk for infants in Africa), giving rise to a stricter legislative regulation of marketing practices. In this regard, we can note the example of the obligation, in France since 2007, for the advertisers of some food products to introduce health information into their communications, such as “For your own health, avoid eating too much fat, too much sugar or too much salt.”

If there have been some opportunist practices on the ground, this should not, in any case, call into question the marketing field as a scientific discipline. From the beginning, marketing has been positioned as a medium for the relationship between a company and its consumers, and the search for well-being remains at the very foundations of the discipline (Andreasen 1994). Today, more than ever, marketing should implement credible and visible actions that will help to increase the legitimacy of companies in relation to the concept of sustainable development (Gabriel 2003). Our society, as a whole, seems to demand marketing that is accountable and demonstrates “its good faith”. This book aims to respond to that demand: it is essential not only to go beyond the marketing-sustainable development split, but also to demonstrate the interest of relying on the tools and analytical frameworks of marketing to serve the cause of sustainability. Marketing and sustainable development can be intertwined in order to achieve a common goal, of strengthening ecological behaviors.

Fourteen chapters, rooted in action and offering rich and detailed views of different topics related to the question posed, are included in this book. The analysis suggested by scholars specializing in the field of responsible consumption is completed by the views of professionals in the field (managing directors, sustainable development heads, managers, consultants, public officials, etc.). These chapters cover five complementary themes.

The first theme explores, in the first two chapters, *the role of the consumer in the green movement.* In Chapter 1, Abdelmajid Amine and Mouna Benhallam consider individuals with a high level of environmental and social concern, who opt for radical changes regarding the traditional consumer pattern. By studying the case of engaged online communities, the authors show that, through their resistance to the market, these consumers are contributing to transforming it by promoting responsible consumption.

In Chapter 2, Mohamed Akli Achabou and Sihem Dekhili, on the other hand, consider the case of consumers less sensitive to ecology. The significant attention paid to the intrinsic characteristics of the offer (such as quality in the case of luxury products) can lead to a rejection of eco-products; some consumers believe that the integration of sustainable attributes decreases the offer's value. More broadly, by highlighting a perceived contradiction between luxury and sustainable development, the authors deduce that the green issue cannot be explored in the same way for different product categories.

The second theme, addressed over three chapters, concerns *waste*, in particular, that produced in the food sector. This is a subject that has played an important role in media and political debates in recent years.

In Chapter 3, Guillaume Le Borgne, Margot Dyen, Géraldine Chaboud and Maxime Sebbane question the role of the consumer in the above-mentioned waste. By analyzing the different structural, contextual and chain organization constraints, the authors call for the responsibility of consumers in waste to be put into perspective, compared to the other actors in the food system. By exploring the case of mass catering companies, they highlight the sources of the ineffectiveness of measures to combat food waste.

In Chapter 4, Amélie Clauzel, Nathalie Guichard and Caroline Riché address this issue in a smaller sphere, that of the family. They outline the role of each family member in the food waste phenomenon and mention the differences in perceptions of food waste within the same household. Moreover, the authors analyze the causes of waste resulting from family purchase and consumption processes.

In Chapter 5, Maud Daniel-Chever, Élisa Monnot, Fanny Reniou and Lucie Sirieix look at a new mode of buying products that is aimed at combating waste: packaging-free products. The authors provide, on the one hand, an understanding of the profiles of consumers buying such products, their motivations and barriers, and on the other hand, an analysis of the offer in this field and the complexity of managing it.

The third theme in this book concerns *communication*. Three chapters are focused on a thorny question in responsible marketing: should companies communicate about sustainability? And if so, how? To what extent can this communication be responsible and effective in relation to the target's expectations? The topic is a particularly important one as the reactions to societal communications are often negative and the phenomenon of *greenwashing* continues to be prevalent.

To respond to these questions, in Chapter 6 Agnès François-Lecompte and Sylvie Foutrel, by considering societal communication as a triptych ("what message", "what channel" and "to whom"), discuss the conditions of its effectiveness. The authors study the question of the credibility of the message and the potential of the societal communication to create value for customers.

This reflection is expanded in Chapter 7 through the contribution of Sihem Dekhili and Samer Elhajjar, which is focused on a particular element of communication: the tone used. More specifically, the authors examine the effectiveness of "provocation" in environmental advertising. Given the consumer skepticism towards the green claims of companies, some brands have opted for greenbashing, a form of communication characterized by sarcasm. The advertisers, in such cases, do not hesitate to deliberately mock green activists and/or trends. The authors detail the motivations that lead brands to favor this communication tone and test its effect on the effectiveness of environmental advertising.

The topic of communication is brought to a close in Chapter 8 by Philippe Odou, Marie Schill and Manu Navarro, who investigate how to communicate effectively on one particular complex topic: climate change. The authors analyze the types of communication and the emotions to stress when targeting adult profiles showing different emotional reactions and intentions to act. At a time of unprecedented mobilization among the younger generation, the authors are interested in the specific case of children, providing a deep understanding of the mental representations of climate change among this target group.

The fourth theme, addressed over three chapters, looks into the issue of *regulation*. Firstly, in Chapter 9, Leila Elgaaied-Gambier and Laurent Bertrandias suggest an interesting analysis of the following question: how can the integration of a marketing approach into public policies for sustainability contribute to making them more acceptable in the eyes of the target audience? The authors discuss different approaches adopted by the public authorities in environmental regulations and discuss their advantages and limitations, before further consideration of the contribution of marketing in terms of strengthening the efficiency of pro-environment public policies.

Then, in Chapter 10, Mickaël Dupré, Patrick Gabriel and Gaëlle Boulbry address the issue of information on the reparability of products. They propose to determine whether this constitutes a selling point, especially among people particularly concerned by sustainability. The authors offer an understanding of the repairability concept and consider the effects of a "reparability label" on consumer perceptions and behaviors in the case of household appliances.

Lastly, in Chapter 11 Mantiaba Coulibaly-Ballet focuses on the fair trade label. Through a study conducted in Côte d'Ivoire, the author questions the role of this cue in both the spread of sustainable production and the encouragement of responsible consumption, in a context where sustainable development is still emerging.

The book concludes with a very current *fifth theme, the digital domain*. Three chapters emphasize the role that technological and digital tools can play in the expansion of the green movement. These tools have the advantage of directly affecting the decisions and behaviors of consumers more than their values, which can lead to improved effectiveness in the adoption of responsible behaviors.

In Chapter 12, Adeline Ochs and Julien Schmitt establish the general framework of the topic by showing the main mobile applications linked to responsible consumption and explaining their mechanisms for influencing consumer behaviors.

Among the different categories of products and services affected by these new digital applications, food products present specificities, as highlighted by Christine Gonzalez, Béatrice Siadou-Martin and Jean-Marc Ferrandi in Chapter 13. The authors look at the issue of the compatibility of digitization and food sustainability. They then establish a range of models that facilitate understanding of how to influence consumer behaviors in favor of responsibility. In addition, the types of digital devices that support the adoption of food sustainability are detailed.

In Chapter 14, Myriam Ertz, Shouheng Sun, Émilie Boily, Gautier Georges Yao Quenum, Kubiat Patrick, Yassine Laghrib, Damien Hallegatte, Julien Bousquet and Imen Latrous go further in examining the benefits of technology 4.0, in terms of extending the lifespan of products and combating planned obsolescence. Through a close analysis of the main features and functions of technologies linked to "Industry 4.0", including additive manufacturing, the Internet of Things, Big Data and artificial intelligence, the authors show how the characteristics of these technologies can help to create augmented sustainable products. This is a major challenge of our century!

If they indicate difficulties in the dissemination of sustainable development (consumer skepticism, limitations of environmental communications, waste, ineffectiveness of environmental measures, etc.), the contributions gathered in this book have the great interest of not stopping at limitations and suggesting concrete recommendations and solutions to improve the effectiveness of organizations. Our work offers evidence of the great potential of the use of marketing tools and approaches to encourage sustainable development. This book should therefore be considered a reformer and a cause for hope!

References

Achabou, M.A., Dekhili, S., Tagbata, D. (2018). Mieux comprendre les difficultés de développement du DOGGY BAG en France : une analyse par l'approche des représentations sociales. *Décisions Marketing*, 92, 53–76.

Andreasen, A.R. (1994). Social marketing: Its definition and domain. *Journal of Public Policy and Marketing*, 13(1), 108–114.

Auger, P. and Devinney, T.M. (2007). Do what consumers say matter? The misalignment of preferences with unconstrained ethical intentions. *Journal of Business Ethics*, 76, 361–83.

Brownlie, D. (2006). Emancipation, epiphany and resistance: On the underimagined and overdetermined in critical marketing. *Journal of Marketing Management*, 22(5/6).

Brundtland, G.H. (1987). Notre avenir à tous. Report, Commission mondiale sur l'environnement et le développement, United Nations, UNEP 021.

Daumas, J.C. (2020). Les métamorphoses de la société de consommation (France, 1945–2020), consommation et histoire. *Carnets de la consommation*, 2, 1–36.

Davies, I.A., Lee, Z., Ahonkhai, I. (2012). Do consumers care about ethical-luxury? *Journal of Business Ethics*, 106, 37–51.

Dekhili, S. and Achabou, M.A. (2015). The influence of the country-of-origin ecological image on ecolabelled product evaluation: An experimental approach to the case of the European ecolabel. *Journal of Business Ethics*, 131(1), 89–106.

Dekhili, S., Coulibaly, M., Dufeu, I. (2017). Les consommateurs perçoivent-ils le prix des produits du commerce équitable comme étant juste ? *Décisions Marketing*, 85, 83–100.

Gabriel, P. (2003). Le marketing comme moyen de légitimation des entreprises dans une perspective de développement durable. *Décisions Marketing*, 29, 67–76.

IFOP (2019). Les Français et l'environnement, vague 1 [Online]. Available at: https://www.ifop.com/publication/les-francais-et-lenvironnement-vague-1/.

Kotler, P. (2011). Reinventing marketing to manage the environmental imperative. *Journal of Marketing*, 75, 132–135.

Thøgersen, J., Haugaard, P., Olesen, A. (2010). Consumer responses to ecolabels. *European Journal of Marketing*, 44(11/12), 1787–1810.

Yannou-Le Bris, G., Serham, H., Duchaine, S., Ferrandi, J.M., Trystram, G. (2019). *Ecodesign and Ecoinnovation in the Food Industries*. ISTE Ltd, London and John Wiley & Sons, New York.

1

Opposing the Market Through Responsible Consumption to Transform It

Abdelmajid AMINE and Mouna BENHALLAM
IRG, University Paris-Est Créteil, France

1.1. Introduction

The increase of ecological and social concerns in recent decades, calling for radical changes to modes of production and consumption, shows the signs of a societal model that is flagging. This model, long propped up by sales techniques that encouraged over-consumption and the unbridled pursuit of possessions and material comforts, is, now more than ever, being called upon to reorient itself around a fairer *sustainable* balance between people, the environment and the economy.

There has been criticism of marketing since the early 1970s (Kassarjian 1971), highlighting its shortcomings and side effects, namely, the harmful incitement to buy, the creation of superficial needs and the devastating exploitation of natural resources. This then developed with time around major concepts, such as "sustainable development" (Nader *et al.* 1971), the "social responsibility of companies" (Manrai *et al.* 1997), consumer opposition (Penãloza and Price 1993) and even theoretical frameworks, such as Transformative Consumer Research (TCR) (Mick *et al.* 2012). The latter supports results-based marketing research on major social issues (health, poverty, the environment, etc.), the results of which are intended to improve individual and collective well-being. All of these changes reflect the gradual dissemination of contestation behavior and discourse to a wider section of the population.

Marketing for Sustainable Development,
coordinated by Sihem DEKHILI. © ISTE Ltd 2021.

This phenomenon is evidence of the normalization of these opposition practices (Amine and Gicquel 2011) and translates into an inclination among consumers to make their behaviors more disciplined, so as to bring about societal changes. In that regard, it is clear that socio-environmental concerns are preeminent, as expressed through the protest-based reactions[1] of consumers (Dubuisson-Quellier 2009). These dissenting movements have given rise to a specific form of opposition of a socially responsible nature (Benhallam 2016)[2], and spark recovery attempts from companies who perceive them as a threat.

With this in mind, and in view of the digitalization of social interactions and the increased speed of virtual exchanges (Kozinets 2012), we propose to explore practices of resistance, particularly those organized by consumer groups online, especially on social networks, given their significant impact.

As an example, following price increases, a boycott campaign against Danone, launched online, took place in North Africa in 2018 and cost the brand 178 million euros[3]. The main lessons that we want to highlight are intended to underline the ability of pressure from consumer movements to force companies to reevaluate their marketing practices and to push the public authorities and associations to rethink the regulatory mechanisms of the markets.

More specifically, we will be working on answering the following two-part question: "How does the online resistance of individuals adhering to a responsible consumption approach express itself in the face of market ideology? And which possibilities for social development does it open up?" In order to do this, a two-year netnographic study was conducted in three virtual communities based around responsible social opposition, specifically, "*Le changement par la consommation* [Change through consumption]", "*Mr Mondialisation* [Mr. Globalization]" and "*Objecteurs de croissance* [Growth Objectors]".

1.2. Corporate adjustment strategies in response to the contestation of market logic

1.2.1. *From an adaptive perspective of uprising recovery by the companies...*

In a fast-moving anti-establishment environment that often generalizes (Dubuisson-Quellier 2009), companies have a tendency to respond to opposition

1 Monnot and Reniou (2013) believe that contestation is a medial element in a continuum, where the starting point would be a feeling of skepticism, conducting an expression of resistance as an end point.

2 SRR: Socially responsible resistance.

3 Available at: https://www.jeuneafrique.com/737259/economie/danone-des-resultats-annuels-en-berne-au-maroc-en-raison-du-boycott-de-2018/.

movements through adaptive and short-term initiatives, such as the provision of products and services that are supposed to be "responsible" (biological, fair, green). These provisions are not necessarily backed up by a thorough review of procedures, means and methods (Koskenniemi 2019). However, companies would have to invest a lot in a transformative policy, having to change their habitual reactive attitude of creating incomplete offers and taking the risk of maintaining opposition, rather than appeasing it. In doing so, companies are only suggesting superficial adjustments based around a traditional sales logic, which hints more at *greenwashing* than true compliance with the values of social responsibility and sustainable development.

As a result, companies that subscribe to the traditional market-oriented system are considered responsible, by anti-consumption groups, for damaging the environment, encouraging over-consumption and spreading poor-quality food (Daniel *et al.* 2008). The one factor these activists have in common is the contestation of a market logic that encourages excessive consumption, without respect for individuals and groups developing practices supporting opposition and anti-consumption, or even deviance (Amine and Gicquel 2011). There appear to be several elements at play at the intersection of different types of socio-environmental contestation, at the micro level, companies and their enslaving mercantile practices, and at the macro level, the overall functioning of society, which is deliberately based on a consumerist logic emerging from the capitalist system.

In fact, even though the market has recovered from some of the criticism, and endogenized it through the development of products labeled "sustainable" or "ethical", the fact of the matter is that significant numbers of citizens remain suspicious and continue their protest and defection from the market (voluntary reduction of consumption, prioritization of local distribution networks, boycotts, etc.). While some consumers approve of companies' efforts to become greener, others continue to contest them or to be skeptical towards them, assuming their intentions to be much less respectable. The latter consider the "socially responsible" branding of products to be a commodification of ethics and an opportunism that only encourages their resistance to such companies. As a result, the socially responsible position demonstrated by companies in the traditional market and seen in the development provisions that are labeled socially responsible have struggled to reach the expected objectives, in particular, because they use the same sales techniques and persuasive registers that have been challenged (Koskenniemi 2019). On the other hand, the failure of these companies to take the socially responsible expectations and principles that consumers associate with sustainable development on board creates a supply of products and services that are considered to be insincere, as they do not seem to align with the values defended by these individuals. Indeed, the various contestation movements, characterized by strong socially responsible concerns, are the manifestation of a persistent criticism of the solely market-oriented offers of companies and of market ideology, despite the attempts made to endogenize them.

With this in mind, traditional marketing, which has, to date, adopted reactive and adaptive strategies towards consumer efforts, should now make a change to become part of a proactive strategy to transform its values, provisions and processes, in order to meet the expectations of socially responsible consumer-citizens.

1.2.2. *...to a transformative market logic under pressure from protest movements*

Further to the socio-environmental criticisms, the market is no longer considered, from a consumer perspective, to be a place where beneficial exchanges can occur. It is instead perceived as a space of influence in which unbalanced exchanges take place, with poorly equipped and influenced consumers who make sub-optimal choices (Dobscha 1998).

The representation of the market from critical sociology (Baudrillard 1970) highlights behavior in opposition to the market, which is considered as combative in view of the market's hegemonic and oppressive role. Moreover, the postmodern paradigm (Firat and Venkatesh 1995) emphasizes the "fragmentation of society" that makes it possible for consumers to free themselves through various alternatives to consumption that constitute as many degrees of freedom. From this point of view, this opposition concerns an "expression of self" more than a "fight". This being the case, the opposition movements sometimes express themselves in relation to the deterioration of the environment, considering the capitalist system that governs the market to be a "structure of destruction" (François-Lecompte and Valette-Florence 2006), and sometimes in relation to the techniques of the market-oriented system, viewing the market as a "structure of domination" (Murray and Ozanne 2009). At the same time, the emphasis is on the "conscientization" of the consumer in regard to the tools and methods of the market-oriented system. Roux (2007) refers to this "awareness" as a precondition for learning about market relationships and then acquiring skills that make it possible to have, *in fine*, a clear idea of the context of trade and the unbalanced relationships of the powers involved.

In order to address the object of criticism of marketing and the market system and/or to articulate the resistance reactions of consumers to this criticism, it is therefore necessary to consider the nature of the market's measures and their socio-environmental impact. As a result, to regenerate, marketing would need to have a genuine paradigmatic break, with the aim of inventing an innovative and deliberately different marketing model (Badot and Cova 2008). The latter should, in particular, identify the expectations and interests of citizen-consumers in advance by measuring, at the ethical level, the impact of their choices and actions on the different stakeholders and, more broadly, on society as a whole. For a company, this proactive attitude should consist of engaging in a thorough process of transformation of methods and practices that integrates more coproduction with the consumer, and researching the meaning of goods offered and ecological and societal concerns.

The communities we have investigated have, in this regard, shown criticisms and denunciations of the market system, as well as guidelines for an alternative marketing model. This position is in line with the values of these opposition communities, whose protest is focused on social and environmental concerns (Benhallam 2016). Indeed, in using the analysis of the data collected within these communities (see Box 1.1), several possibilities for change, or even for transformation, present themselves to the field of marketing.

In order to respond to our question, and to be able to understand how the communities in opposition to market ideology express themselves and which possibilities they can open up in terms of marketing practices, we have chosen a qualitative netnographic approach. We are interested in three distinct virtual communities who share their denunciation and rejection of market-oriented ideology (and its corollaries, the capitalist system and excessive consumption), while showing their stance in favor of sustainable and managed consumption.

Our approach relies primarily on the study of one main community, "*Le changement par la consommation*", and two secondary communities, "*Mr Mondialisation*" and "*Objecteurs de croissance*". The selection was based on the stated position of the community, the high number of members, regular posting by the central moderator(s) and significant member activity. The inclusion of the secondary communities aimed to allow the comparison of the results from the main community with those of other fields, in order to ensure their relevance and to gain external validation.

The netnographic study was carried out through a long (24 months) immersion in the three communities with non-participative observation, ensuring a non-intrusive environment with natural interactions that preserves authenticity on the ground and the reliability of the data (Hewer and Brownlie 2007). However, in order to respect the principles of ethics (Kozinets 2012), the central moderator(s) were asked questions relating to the use and publication of data through private messages, ensuring transparency in regard to our intentions and the use, for purely scientific purposes, of the information collected.

The significant body of information available was filtered through the use of key words (namely, resistance, opposition, rejection, (over)consumption, consumerism, commodification, consumer ideology/values, etc.) and through the calculation of a Buzz Indicator (BI)[4], developed on an *ad hoc* basis to select the most salient and relevant data. We thereby retained the subjects and concerns that are characterized by both strong engagement (BI score) and significant recurrence within each of these communities. The stable results obtained in this way have been put into a reference framework suited to the opposition phenomenon, the interpretation of which enables the proposition, *in fine*, of possible changes to marketing practices.

Box 1.1. *Netnographic study of three online communities involved in responsible consumption*

Communities	Aim of the community	Creation date	Number of central moderators
Le changement par la consommation	Challenge the capitalist system and the modes of consumption it creates and propose alternative solutions. The different community posts are intended to make its members react, while allowing information to circulate within the social network to reach a larger audience.	May 2011	1 moderator
Mr Mondialisation	Provoke reflection through information, videos and pictures of issues related to the dominant system and its impact on the life of humankind and the environment. The aim is to lead its members to debate and create ideas and solutions. The community regularly organizes protest activities online and on the ground.	November 2008	Several moderators
Objecteurs de croissance	Defend and spread the school of thought that "degrowth does not mean living less, but better with more goods and more connections". Involvement in debate and interactions between members are a requirement. Emphasis is also placed on reflection and raising awareness, rather than on proposing solutions.	June 2012	1 moderator

Table 1.1. *Main features of the three communities studied (source: (Benhallam 2016))*

4 The Buzz Indicator (BI) refers to the degree of involvement of the community members with a given topic, suggested by the breadth of participation around it (likes, comments and shares). In order to be selected, the BI must be higher than the average of the BIs for a "t" period (from the 1st to the 30th of each month).

The three communities studied (see Table 1.1), both the central moderators and active members, aim to challenge the dominant modes of consumption through an ongoing effort to deconstruct the conventional market-oriented system, with the aim of proposing alternative adjustments, or even breakthrough solutions.

The analysis of thematic content through exchanges between the members of the three community sites has enabled, after the clean-up of discussions, the data to be structured into two meta-topics reflecting the main forms of expression of the opposition of these movements. These meta-topics are divided into four categories covering the challenging of the dominant ideology, the re-establishment of trust in trade relationships, the reconsideration of product offers and the reconfiguration of supply and distribution networks.

1.3. Ideological and institutional categories of expressions of contestation

1.3.1. *Towards a redesign of the dominant ideology of the market system*

The strong denunciations by the communities of the dominant way of thinking are intended to produce an ideological opposition, in order to rebuild or reshape the ideas and actions of their members and beyond. This ongoing work of re-ideologization is aimed at producing and then establishing new values in opposition to the market-oriented view: the primacy of solidarity, trade, being free of charge, preservation of health, respect for nature and freedom of choice. The opposition of the two value systems is explicitly highlighted in the posts made by members: *the quest to preserve life in all its forms in order to pass it on to future generations* is in contrast to *the structural trend of exhausting resources for insatiable financial gain*. The confrontation of these two world views champions an alternative social model, defended by the three communities and supporting a well reasoned consumption that is conscious of its effects on the environment, health and well-being.

In addition, and unanimously in the three communities studied, the members place themselves in opposition to the dominant system by stigmatizing the relentless pursuit of profit and the race to accumulate material goods, on the one hand, and by praising slowness and sobriety (Rabhi 2010) through reasoned consumption for human and environmental well-being, on the other.

However, despite strong criticisms of marketing, it should be noted that the three communities are not opposed to the idea of using marketing's weapons against it:

developing logos, visual identities and slogans; and monitoring posts to recruit more members, respond to skeptics, and increase the fame of and loyalty to the community site. For example, the "*Objecteurs de croissance*" community, whose name is indicative, immediately suggests an anti-growth message. It has also developed a visual identity with a snail logo, in order to evoke the idea of slowness that the community lauds to promote a degrowth process. The visual identity of the community also has the slogan "Fewer goods, more links" (in the original French, the rhyming "*Moins de biens, Plus de liens*"), to establish the anti-growth idea and underline the importance of social connections.

1.3.2. *Towards reestablishing a relationship of trust with the consumer*

In underestimating the interest and concerns of consumers, and society more generally, in terms of preservation of health, respect for the environment and the struggle for equality, companies have long allowed themselves to adopt dominant positions towards consumers. This is why the relationship with business is perceived, by the three communities studied, as a relationship always defined by power, where the citizen-consumers are attempting to destroy the "totalitarianism" of the market system (Baudrillard 1970) and its powers of domination (Murray and Ozanne 2009), as well as methods that are viewed as "perverse" (Penaloza and Price 1993). The result of the hegemony of the company over consumers is that the company has long dealt with features characterized by their dissipation and their inertia, allowing it to exercise its influence over them. On the other hand, what has changed is that, through consumers grouping together in collectives, communities or associations, they have been able to recalibrate the power relationship, first initiating and then imposing alternative solutions to the traditional market to counter this hegemony. This being the case, the community members confirm that there should be freedom from this dependence by inventing other forms of consuming that fall outside the scope of the traditional market. This view translates into a trend towards consumer empowerment, by imparting a world view that is more conscious of health, more respectful of the environment and more engaged in combating waste. The following statement, issued by the "*Le changement par la consommation*" community, illustrates the anti-waste idea by promoting self-consumption in households: "Heyyy… I went over to homemade a while ago, so satisfying to create your own products (treatments, make up, household products, candy…) A great site for that: http://aroma-zone.com/aroma/accueil_fra.asp with everything, loads of instructions…" (www.facebook.com/le-changement-par-la-consommation-218263484822051/).

As a result, the domination approach, questioned at the individual and group levels by consumers, is challenged by a collaborative and socially responsible approach, where companies and citizens cooperate on a shared project, founded around basic values that must be respected. This project, balanced between human, environmental and economic components, would be the basis for the development of suitable provisions and measures to be implemented to the benefit of the different stakeholders. This possibility for cooperation between companies and consumers could be a credible alternative to a conventional marketing approach that is not considered socially responsible. As well as its inclination to create a new, fair and balanced relationship between the stakeholders, this approach would constitute a means for the potential reconciliation of companies with the protest movements that target them, e.g. the anti-consumers, protesting citizens, counter-cultural communities[5] and communities of socially engaged individuals, with a view to encouraging business to take on a new idea of the world.

More broadly, for the online communities studied, this concerns the reconsideration of the field of action and the use of marketing, which should not be limited to sales but should take into account the impact of its choices on health, resources and the environment. This means a shift in the center of gravity of power, from the sole interest of the company to an integration of the collective interest. To that end, research highlights that brands will regain the trust of consumers when they show their ability to assume civic responsibilities (Dubuisson-Quellier 2009) alongside their role as an economic actor. Consumers and civil society will make their presence felt as stakeholders in considerations, positions and economic and social choices, as basic elements of an alternative and inclusive vision of the market.

1.4. Pragmatic and operational categories of market contestation

1.4.1. *Towards a sustainable reconsideration of product offerings*

Stemming from the values defended by the members of these communities involved in ecologically responsible consumption, the offering of products and services is supposed to respond to a three-fold demand: that they are useful, sustainable and recyclable. According to the members of these communities, this is not only about promoting a "really useful" product, but also one with sustainable and recyclable properties, with a view to limiting the depletion of natural resources. The "sustainable" nature stands out as one of the features that the members of the three communities expect most, as evidenced by, among other things, the tendency to avoid impulsive purchases of short-lived and useless "gadgets".

---

5 Individuals who are part of a social group consciously and deliberately opposed to a dominant culture through individual or collective actions of varyling levels of visibility.

Product offerings and practices concerned	Expectations in terms of the suppression of superfluous or unsustainable products	Expectations in terms of the development of responsible and sustainable products	Inherent or mobilized values
Cosmetic, hygiene and self-care products	Industrial and over-packaged products	Natural products based around natural components and with natural packaging	Protection of health and the environment
Food products	Products transformed by and produced by traditional agriculture	Products from bio-agriculture and/or small farmers and green cooperatives, with as little modification as possible	Protection of health and the environment, solidarity
Pharmaceutical products	Chemical products with sometimes serious side effects	Natural products with proven benefits for people and the environment and without notable side effects	Protection of health and the environment
Motor vehicles and technological products	Polluting products subject to planned obsolescence	Environmentally-friendly substitute products: bikes, electric cars, public transport with less of an impact on the environment	Protection of the environment
Distribution channels	Widespread distribution and traditional networks	Networks for swaps, trades and giveaways	Solidarity, *de*commodification of trade
		Circular economy (second-hand shops and purchases, recycling)	Reduction of waste
		Short "small producer" networks and/or specialized structures	Protection of health and the environment, reduction of waste
Consumption-related events	Sales periods, periods with promotions and special events (encouraging over-consumption)	Days/weeks without purchases Encouragement of minimalist and activist purchases "useful" to people and the environment	(Re)developing social connections, the primal importance of the usefulness of the object, reduction of waste
Communication	Aggressive and intrusive advertising, advocating the accumulation of goods, possessions and appearances	Social advertising promoting the values of responsibility, protection of the environment and respect for human health	Promoting individual and collective well-being, protection of health and the environment

Table 1.2. *Views of the development of marketing practices through the lens of responsible consumption*

In the same vein, these communities seem to be particularly attached to combating planned obsolescence, which is both a tool facilitating the continuation of the traditional market system (limiting the lifespan of products in order to encourage consumption) and a factor in the deterioration of the environment (promoting the waste of resources and the creation of refuse). Beyond the fight against the waste created by planned obsolescence, the community members also mobilize against the producer practices tainted by "irresponsibility", "manipulation" and "cynicism". In this regard, they propose strategies to circumvent planned obsolescence through mechanisms to repair used or malfunctioning objects (clothes, household items, cars, furniture, etc.), including by sharing sites advocating personal repair: "Here's a link that you should like! The dream for combating the planned obsolescence of products: www.commentreparer.com. Don't throw away, repair!" (Change through consumption).

Lastly, in contrast to the rejection of unsustainable products and services triggered by the traditional market, the members of the three communities opt for offerings that are compatible with the values of responsible consumption. This attitude paves the way for a large range of new behaviors suited to well-reasoned consumption. It is precisely at this level of analysis that researchers and decision-makers must add their will to act, adapt and even to reform, as advised in Table 1.2. This shows two views of the development of product offers at the dawn of sustainable consumption by considering, on the one hand, the expectations of consumers involved in suppressing non-compliant products (those harmful to people and their environment) and, on the other hand, the expectations in terms of developing new products characterized by usefulness, sustainability and responsibility.

1.4.2. *Towards a necessary reconfiguration of supply and distribution channels*

In the same vein as the obligation to rethink product offers, the members of the communities studied take into account their search for alternative supply structures, which are deployed alongside traditional distribution networks, which are characterized by unethical practices, namely, an appropriation of value and an unequal distribution of profits. This naturally leads to a shift towards substitute supply networks that respond to the imperatives of responsibility and sustainability, such as short, circular or cooperative networks, among others. These alternative structures rely on different operations and aims than those dependent on the convention model, insofar as they respond to expectations for healthy and authentic products, created in decent working conditions, for a fair price and with less of an environmental footprint.

While it is well known that these alternative structures already operate alongside the market, they are the subject of repeated demands from the members of the three communities, who call for them to become widespread throughout society as a whole and to progressively become the norm in consumption. According to these communities, short networks that use a smaller number of intermediaries should be prioritized, thereby promoting a more equitable distribution of revenues among (small) producers and retailers, and the proposition of more accessible prices for end consumers. To that end, a number of members have attributed the high cost of "green" products to the currently significant number of intermediaries between the production stage and the product becoming available to the end consumer. The supply networks promised by these communities are therefore characterized by direct purchasing from farmers, local producers, peasant associations, green cooperatives and activist structures, such as Enercoop and Kokopelli. To that end, the members share web links to various shopping websites to encourage, in particular, the use of Amap (*Associations pour le maintien d'une agriculture paysanne* – Associations for the Maintenance of Peasant Agriculture), as highlighted in the following post: "The Amap directory for France: www.reseau-amap.org. On this site you will find: information on what an Amap is, an Amap directory and a guide to help you to create your own Amap" (*Le changement par la consommation*). Lastly, the members of these communities promote experimentation with new forms of trade, following the example of networks of mutual aid, second-hand sales and free provision of goods and services between individuals sharing the same values of responsible consumption through the organization of regular or periodic events (such as trade fairs, markets or swap sites). A number of sites have thereby been made available to members, who have many spaces dedicated to talking about their experiences and recommendations from their participation in these events. In doing so, they contribute to creating a sort of library of knowledge that is distributed between the different members and continually enriched by the experiences shared. The example of the site recup.net, shared on several occasions by the community moderator of "*Le changement par la consommation*", is telling. It highlights the importance of the circular economy in reducing the production of waste, protecting the environment and combating planned obsolescence:

> The Recup site is a site for donation, reclamation and free swaps, where everyone is able to propose and give over the Internet, rather than throwing out what they want to get rid of. Recup.net is not a site

> facilitating sales or trades [...], but a site for donations. It must remain a sharing space where anyone can find free recovered objects.[6]

1.5. Conclusion and implications

The disenchantment and discontent expressed towards the consumerist society by socially responsible consumers organized in online communities cannot be reduced to a circumstantial phenomenon, but falls within a radical questioning of the traditional consumer model (Sansaloni 2006). This kind of resistance through engagement with and for a socially responsible consumption is thereby characterized by the structural and collective dimension of the phenomenon and by its appearance in different forms (ideological, institutional and operational) of consumption. In fact, numerous experts agree that in terms of a "citizen", "socially responsible" or "sustainable development" position, it is not a question of a temporary enthusiasm, but rather of a lasting change in the collective consciousness and the practices of individuals. The serious trends that are taking shape and establish responsible and sustainable consumption as viable and fast-growing methods are evidence of this appetite for the values of reasonable consumption, environmental protection, solidarity and social justice. The account in Box 1.2 echoes this and shows the main ideas of an expert in socially responsible consumer engagement.

However, in order to be consistent and effective, this critical view of consumption should be accompanied by a reconsideration of the analytical frameworks used, as they currently rely on the same logics and measures as the market. Indeed, following the position defended by Koskenniemi (2019), we think that, in its current configuration, the opposition between "normal" or usual consumption and socially aware consumption is overrated and tends to reify the duality of these two concepts by establishing it as self-evident. However, the normalization of this duality goes against the critical approach that must be used in a relativist view, as what is currently considered oppositional or deviant (here) will not or will no longer necessarily be tomorrow (elsewhere). The typical illustration of this is in the consumption of organic products, long presented as a practice in opposition to the consumption of conventional products, but now considered entirely normal since it has been taken up by the market. This change from marginalization (opposition) to normality is typical of a swing movement that affects the norm and therefore deviance (Amine and Gicquel 2011). As a result, there is neither normality nor deviance nor resistance in absolute terms; everything is a question of context

6 Available at: www.recup.net, "Providing a space for getting rid".

and the prevailing interpretative framework in a social group in a given space and at a given time.

The citizen-consumers involved in sustainable consumption have a key role in changing business practices. Beyond the mere act of buying, it is their role as "influencers", particularly suited to social networks, that will have a greater impact on these practices. This role can largely be broken down as follows:

– *provide information to independent experts on the problems and solutions, and share all of the elements collected.* There are many falsehoods about responsible consumption that put companies in delicate situations. Some solutions that seem good, proposed by influencers on social networks, are sometimes bad at the scientific and environmental levels. For example, in textiles, plant dyes and natural materials are in particular demand, while numerous studies have concluded that these dyes are more polluting than synthetic ones, as they require many phytosanitary products. The engaged consumer should make an effort to get their information from a range of independent sources;

– *become spokespeople for good solutions, including among consumers with different opinions.* Engagement is also making the effort to compare one's ideas with those of consumers with different opinions on different platforms. A true responsible product, from the moment it is useful, is the widespread "core product", made in a sustainable manner. It is not the current optimum sustainable offering alone that will reduce our impact on the planet. In order not to encourage companies to develop separate "sustainable" offerings, but to transform their "core" offerings into sustainable offerings, the engaged consumer aims to convince the unengaged consumer, as it is together that they will change these practices;

– *sign petitions to support draft legislation based on good sense.* In ultracompetitive markets, such as the food and textile markets, a brand that works alone has little chance of making the market change quickly and would be attempting economic suicide if it decided to withdraw its products or make them more sustainable. Brands are prisoners of a system and have a competitiveness that puts a stop to green ambitions. The suppression of single-use plastics works because it is a law that imposes the same rules on everyone. The engaged citizen-consumer can also make their voice heard by signing petitions launched by civil society to support draft legislation and bring about change in the market as a whole.

Box 1.2. *Interview with Thomas Radal, European representative of Ulula and a freelance consultant on responsible sourcing*

As a result, by introducing meaning to consumption to respond to greater social values, the socially responsible engagement of consumers and online communities seems to play a disruptive role while invigorating the market-oriented system. This phenomenon is as susceptible to having a "regulatory" impact on the market system by disciplining its by-products, as it is to triggering a reconsideration of the unsustainable behaviors of individuals and organizations. It is also a catalyst of responsible innovations in both marketing strategies and the redesign of analytical frameworks and tools to support its development, structuring and regulation.

On the other hand, there are still a number of difficulties that hinder the generalization of sustainable consumption practices, owing to their cost (the price of green products), their limited accessibility (transfer from the producer) and a lack of related information (the benefits of these products, tips for getting them for a lower price and less effort). These restrictions constitute avenues for improvement and fields of research with a view to making these alternative modes of consumption accessible to a greater number of people. With this in mind, the role of both the market sector (producers and retailers) and, especially, the non-market sector (public authorities and related fields) is crucial in imagining and implementing creative solutions to normalize these consumption practices.

In fact, the slowness of the generalization of these responsible consumption practices and the accessibility of these offerings risk keeping this movement as a fringe one, and even creating an attitude renouncing behavioral changes and opposition to the market-oriented ideology. The main reasons for this are linked to the clear imbalance of power (consumers versus the market system) that feeds doubts as to the effectiveness of socially responsible protest actions. Indeed, including the members of the community groups, less involved people remain dubious of the outcomes of their demands in terms of effective impact on the environment and on society, leading to periodic defections within the ranks.

1.6. References

Amine, A. and Gicquel, Y. (2011). Rethinking resistance and anti-consumption behaviours in the light of the concept of deviance. *European Journal of Marketing*, 45(11/12), 1809–1819.

Badot, O. and Cova, B. (2008). The myopia of new marketing panaceas: The case for rebuilding our discipline. *Journal of Marketing Management*, 24(1/2), 205–219.

Baudrillard, J. (1970). *La société de consommation*. Denoël, Paris.

Benhallam, M. (2016). La résistance socialement responsable à la consommation : proposition de conceptualisation et implications pour les pratiques du marketing. PhD thesis, Université Paris-Est Créteil, Créteil.

Daniel, M., Siriex, L., Bricas, N., Figuié, M. (2008). Du choix des produits aux styles de vie : un repérage élargi des modalités de résistance des individus pour un développement durable. In *Colloque consommation et résistance*. Université Paris-Est Créteil, Créteil.

Dobscha, S. (1998). The lived experience of consumer rebellion against marketing. *Advances in Consumer Research*, 25, 91–97.

Dubuisson-Quellier, S. (2009). *La consommation engagée*. Presses de la Fondation nationale des sciences politiques, Paris.

Firat, A.F. and Venkatesh, A. (1995). Liberatory postmodernism and the reenchantment of consumption. *Journal of Consumer Research*, 22(3), 239–267.

François-Lecompte, A. and Valette-Florence, P. (2006). Mieux connaître le consommateur socialement responsable. *Décisions Marketing*, 41, 67–79.

Hewer, P. and Brownlie, D. (2007). Cultures of consumption of car aficionados: Aesthetics and consumption communities. *International Journal of Sociology and Social Policy*, 27(3/4), 106–119.

Kassarjian, H.H. (1971). Incorporating ecology into marketing strategy: The case of air pollution. *Journal of Marketing*, 35(3), 61–65.

Koskenniemi, A. (2019). Deviant consumption meets consumption-as-usual: The construction of deviance and normality within consumer research. *Journal of Consumer Culture*, September 12.

Kozinets, R.V. (2012). *Netnography: Doing Ethnographic Research Online*. Sage, London.

Manrai, L.A., Manrai, A.K., Lascu, D.N., Ryans Jr, J.K. (1997). How green-claim strength and country disposition affect product evaluation and company image. *Psychology & Marketing*, 14(5), 511–537.

Marion, G. (2003). Idéologie et dynamique du marketing : quelles responsabilités. *Décisions Marketing*, 31, 49–61.

Mick, D.G., Pettigrew, S., Pechmann, C., Ozanne, J.L. (2012). *Transformative Consumer Research for Personal and Collective Well-Being*. Routledge, New York.

Monnot, E. and Reniou, F. (2013). "Ras le bol d'entendre parler d'écologie !" : comprendre la contestation des discours écologiques par les consommateurs. *Décisions Marketing*, 71, 93–109.

Murray, J.B. and Ozanne, J.L. (2009). The critical participant. *Journal of Marketing Management*, 25(7/8), 835–841.

Nader, R., Ross, D., English, B., Highland, J. (1971). *Action for a Change. A Student's Manual for Public Interest Organizing*. Grossman Publishers, New York.

Penaloza, L. and Price, L. (1993). Consumer resistance: A conceptual overview. *Advances in Consumer Research*, 20, 23–128.

Rabhi, P. (2010). *Vers la sobriété heureuse*. Actes Sud, Paris.

Roux, D. (2007). La résistance du consommateur : proposition d'un cadre d'analyse. *Recherche et applications en marketing*, 22(4), 59–79.

Sansaloni, R. (2006). *Le non-consommateur*. Dunod, Paris.

2

Luxury and Sustainable Development: Companies and the Challenge of Overcoming Consumer Reluctance

Mohamed Akli ACHABOU[1] and Sihem DEKHILI[2]

[1] *IPAG Business School, Paris, France*

[2] *CNRS – BETA, University of Strasbourg, France*

2.1. Introduction

Can the luxury sector still avoid its environmental and social obligations to society at a time when all other sectors of activity are facing increasing pressure from various stakeholders to tackle the challenges of sustainable development? In addition to strengthened environmental regulations and strong activism from non-governmental organizations (NGOs), consumers have shown a deep concern for societal issues.

It is obvious that, in this context, luxury companies must follow this movement at the risk of receiving blame; they have long been scrutinized because of their significant economic growth. Moreover, this sector has often been the subject of strong criticism linked to social issues. For example, the salespeople at a Gucci store in Shenzhen, China, denounced inhumane working conditions in 2011. The film *Blood Diamond*, for its part, revealed to the public the impact of the illegal trade of precious stones (destined for Western luxury businesses), in exchange for weapons, in terms of inter-ethnic massacres, notably in Sierra Leone and Liberia.

Marketing for Sustainable Development,
coordinated by Sihem DEKHILI.

These controversies around the practices of luxury companies present a significant risk to brand image. A commitment to sustainability could mitigate this risk and be a differentiating factor on the market, especially as consumers have expressed new expectations of product quality that take sustainability into account.

More than that, the luxury industry should assume a leadership position in terms of a commitment to sustainability (Davies *et al.*, 2012). Firstly, it acts as a model for other sectors to follow, as it is often a source of creativity. Secondly, the luxury industry has significant financial resources that would allow it to make responsible investments.

However, if luxury companies choose to engage in responsible activities, they are confronted with the difficulty of convincing others of the suitability of this strategic choice. Perceived contradictions between the values of sustainable development (sharing, donations, second-hand goods, recycling, etc.) and those of luxury (exclusivity, hedonism, ostentatiousness, etc.) can create a dissonance in the minds of consumers (Dekhili and Achabou 2016). In addition, responsible practices (the use of recycling and vegetable materials, etc.) can have a negative impact on the perceived quality; luxury products have historically been associated with the use of high-quality raw materials (Achabou and Dekhili 2013; Dekhili *et al.* 2019).

The aim of this chapter is to shed light on the issues related to the luxury industry's commitment to sustainable development. In particular, it proposes to highlight the actions that marketing can offer luxury companies to bring together two notions that may seem contradictory: luxury and sustainable development.

Our contribution will be structured around two main sections. First, we will discuss the strategic interest for luxury brands in integrating sustainable development issues, as well as the risks associated with this choice. Second, we will detail the sources of the perceived contradiction between luxury and sustainable development, and will present solutions to overcome it. For illustrative puposes, an interview conducted with the founder of an innovative luxury start-up, Just Ananas, which supports animal welfare, is included.

2.2. The commitment of the luxury sector to sustainability: an unavoidable but risky strategic choice!

If, in the past, the luxury industry has made headlines for its double-digit growth, even in times of crisis, today it is its commitment to sustainable development that is at the heart of the debate. The question is no longer whether or not this sector should integrate environmental and social issues, but rather how to do so while safeguarding its specificities, notably the exceptional quality of its products.

2.2.1. *From luxury that wastes natural resources to "sustainable luxury"*

If considering the history of luxury, it is possible to go back as far as the time of the pharaohs. During this period, luxury was reserved for the pharaohs, their spouses and some high-ranking dignitaries (embalming the body, building tombs, etc.). In the West, luxury products were created by traditional craftspeople and in very limited quantities, destined for the wealthiest individuals. Since the Second World War, a trend towards the industrialisation of luxury production has been observed. It aimed to respond to the growing demand of populations frustrated by years of privation (wars) and whose quality of life had improved. The luxury industry progressively underwent a strong concentration with the emergence of large groups, such as LVMH, today's world-leading luxury group.

However, despite the fact that the emergence of this concept goes back several years there is no consensus on its definition among the academic community (De Barnier *et al.* 2008). When summarizing the features that can describe it, there are four that stand out. The first, naturally, is quality. A luxury product or service must necessarily be of impeccable quality. Superior quality is associated above all with the use of so-called noble and prestigious raw materials. These materials must be worked with a specific *savoir-faire*, by an artisan (real luxury), and if a machine is involved it must only be in a small part of the production process. The superior quality confers another important characteristic on the product, namely longevity. Luxury products are known for their ability to last a long time. Contrary to the "fast fashion" industry that adopts a logic of short-termism and planned obsolescence, luxury is managed with a long-term perspective (Janssen *et al.* 2014). In this regard, the after-sale services of brands increase the length of the products' lifespan. For example, the Louis Vuitton brand has five repair centers in Japan that ensure that damaged products can continue to be used. The third and fourth characteristics of a luxury item are scarcity and exclusivity. Scarcity can be the consequence of natural or technological constraints (innovations) or can stem from constructed constraints (virtual rarity) (Catry 2007). Exclusivity, for its part, reflects the fact that the purchase of luxury products remains the reserve of a very limited segment of the population. It is also linked to the concepts of exclusion ("I am the only person to have the product") and stratification of the population. For Vickers and Renan (2003), scarcity and exclusivity are important for luxury consumers who are seeking to show their membership of a group or confirm their social status.

To these four essential characteristics of luxury, a new dimension, which has become unavoidable, has recently been added: sustainability, leading to the concept of "sustainable luxury". Luxury consumers have now broadened their expectations of quality to include environmental and social attributes. Through this choice, they can display a positive image, of being responsible people, and thereby compensate for the negative values often associated with luxury, namely ego, superfluity, irrationality and excess (Cervellon 2013; Janssen *et al.* 2014).

Surprisingly, even if studies have recently attempted to explore the possible links between the two notions of luxury and sustainable development, to our knowledge, there is not yet a consensual definition of the concept of "sustainable luxury". In this chapter, we propose to define it as: "a product or service resulting from a production process that makes it possible to obtain an irreproachable quality in the respect of human and animal welfare, and whose lifecycle generates reduced impacts on the environment". The luxury industry has never been as scrutinized in terms of the environmental and social consequences of its practices as it is today. The number of articles in the press implicating luxury businesses in scandals have noticeably increased. For example, Gucci has been targeted for the inhumane working conditions in its retail outlets in China, and Burberry for the deliberate destruction of its unsold inventory in order to preserve the rarity of its products.

In the face of these criticisms, the integration of sustainable development principles in the luxury industry has become unavoidable. In this regard, two strategic trends have been observed. A few rare companies have decided to make this new challenge central to their business model. The Stella McCartney brand is an interesting example of this; it has, for instance, banned all animal raw materials from its workshops. Other companies have chosen to limit the inclusion of sustainability to certain stages of their value chains. Table 2.1 provides some examples of responsible initiatives taken by luxury brands in recent years.

Nature of engagement	Examples of brands	Examples of actions taken
Use of responsible materials	Hermès	Petit H: recycling leather and silk offcuts to produce small accessories
	Gucci	"Green" shoe soles
	Saint Laurent	New Vintage: products made by recovering unused fabrics from previous collections
Labeling	Hôtel Fouquet's Barrière	Obtaining ISO 9001 (quality), ISO 14001 (environment) and SA 8000 (social) certifications
	Le Grand Hôtel Cannes	Green Globe Label
Creation of associations	Cartier	Founder member of the Responsible Jewellery Council

Table 2.1. *Examples of responsible initiatives in luxury industry*

2.2.2. *Luxury companies and the challenge of sustainability*

The various responsible initiatives taken by luxury brands have been the subject of communications, whether in detailed corporate social responsibility (CSR) reports or on their own websites. This new trend for transparency, contrary to the policy of secrecy that has always characterized the luxury sector, is intended to respond to the different pressures coming from consumers and other stakeholders. However, they expose luxury companies to two major risks. The first is being accused of greenwashing. The green market is indeed characterized by the multiplication of companies' opportunistic behaviors that have contributed to the increase of consumer skepticism towards environmental information. In the luxury sector, Kering has recently been accused of greenwashing; the group, which considers sustainable development one of its strategic priority areas, has received criticism for its treatment of animals, especially rabbits, whose skins are destined to be used in the production of the brand's coats. This situation refers to the concept of "hypocrisy" mentioned by DeTienne and Lee (2005). The dissemination of ecological information is, in fact, not enough to build a responsible image because companies may act inconsistently in regard to their promises of environmental and social engagement. A perception of hypocrisy has a negative impact on consumer beliefs about a company's responsible commitment (Wagner *et al.* 2009). When the brand is perceived as "showing off", owing to the misuse of the environmental argument, consumers doubt the legitimacy and authenticity of its claims (Cervellon and Shammas 2013). Luxury companies must thus be convincing in the credibility[1] of their commitment to sustainable development and in the sincerity of their communications on this issue.

The second risk to which luxury companies are exposed is when their engagement relates to a negative perception of their new responsible offer. Environmental actions, such as recycling or even social initiatives, such as the replacement of animal raw materials with alternative ones (synthetic or of vegetable origin), can in fact have a negative impact on the perceived quality of luxury products.

2.2.2.1. *Environmental practices that could damage the perceived quality of luxury products*

Luxury brands such as LVMH and Estée Lauder have implemented actions to optimize water and energy consumption by encouraging the use of renewable

1 Credibility has two important dimensions: trustworthiness and competence. Trustworthiness consists of the receiver of the message trusting that the source is putting out honest information with no ulterior motive. Competence refers to the perceived experience of the source (Eisend 2002).

resources. Others have adopted recycling, such as Yves Saint Laurent (now Saint Laurent) which, in June 2009, launched its New Vintage collection, created with unused fabric from previous seasons, and Hermès, which in its Petit H workshops produced small accessories made from offcuts of leather or fabrics, and even 16ᵉ Sud, which makes bags from recycled leather.

Some brands are convinced of the positive impact of environmentally-friendly practices on the perceived quality of their products. Kering, for example, considers sustainability to be an extension of quality; the natural dimension that characterizes environmentally-friendly materials can strengthen the superior quality of luxury products. In reality, the final outcome depends on the judgment of consumers, who can be skeptical of the inclusion of sustainable materials in luxury products (Achabou and Dekhili 2013). Luxury companies should demonstrate caution in the selection of new sustainable raw materials and plenty of creativity in how they are used.

As an example, Gucci's communication (via its website) on its environmentally-friendly shoes emphasizes a production process that brings together creativity and responsibility, with the ongoing concern of responding to consumer expectations in terms of quality, unique design and so on:

> Gucci is delighted to announce the launch of environmentally friendly soles made from bioplastics (biodegradable materials) [...] This new project is emblematic of the aim of the House, which is to interpret the desires of the modern consumer in a responsible way though sustainable items, while maintaining the balance between the timeless values of style and superior quality and an expanding ecological vision.

2.2.2.2. *A social concern harmful to the perceived quality of luxury products*

One of the issues that has been at the heart of debates on the luxury industry for a number of years is animal welfare. In addition to various NGOs that campaign for animal rights (PETA, L214, etc.), we have seen the emergence of a consumer movement (veganism) which considers that any use of animals is unethical, that humans have no right to use animals and that animals should be considered living species just like humans (Plannthin 2016, pp. 49–122). In this pressured context, a range of strategic responses from luxury companies has been observed. Some brands, such as Lagerfeld, have decided to continue to use animal raw materials, thereby ignoring the criticisms of animal rights advocates. Others, such as LVMH and Hermès, have opted to buy farms in order to secure their supplies and have better control over livestock conditions. Still others, such as Chanel, have preferred to limit the use of animal skins to those of animals destined to end up as human food. Lastly, another group of brands has made the radical choice of ceasing to use animal raw materials and to replace them with substitutes (see Box 2.1). This movement was started by the Calvin Klein brand in 1994 and has since been adopted by a number of other companies; Stella McCartney (2001), Ralph Lauren (2006), Vivienne Westwood and Tommy Hilfiger (2007), and, since 2016, Michael Kors, Jimmy Choo, Donna Karan, Maison Margiela, Giorgio Armani, Hugo Boss, Donatella Versace, Burberry and Gucci.

While the decision to include alternative materials has been welcomed by animal rights groups, the same cannot be said of fur producers. The president of the French Fur Trade Federation saw this as a communications campaign.

However, beyond the pro- and anti-fur debate, it is the consumer response that is the biggest challenge for luxury companies. The choice to replace animal raw materials with vegetable or synthetic substitutes comes back to affect an important aspect of luxury, namely quality. A recent study conducted in the French and Saudi contexts (Dekhili *et al.* 2019) sheds light on this subject. The consideration of the challenge of animal welfare leading to the use of vegetable leather has a neutral effect on the perception of luxury products among French people. However, the Saudis tend to lower their assessment of the quality of luxury products made from vegetable leather. This illustrates the ineffectiveness of the animal welfare argument among luxury consumers and highlights the influence of culture on the perception of the quality of sustainable luxury products. Beyond the low environmental and social awareness of Saudi consumers, the social pressure in a cultural context based on appearance can explain the reluctance observed towards the sustainable offer. Some brands try to overcome this risk by presenting the alternative materials used as innovative and luxurious. This is the case of Stella McCartney, for example, which, on its website, associates vegetable leather with innovation and animal leather with an old-fashioned material: "Scientific research is discovering new mixed fibers that are therefore becoming rare products, which is why innovative materials must be perceived as a real luxury, more so than leather, which has become a relatively common product today".

Today, the issue of animal welfare is at the heart of the debate within the luxury industry. For example, under pressure from various stakeholders, including NGOs, a number of companies have made the choice to abandon animal fur and to replace it with alternative raw materials. Some have opted for synthetic fur, a petrol by-product, despite the fact that they are criticized for the pollution this causes. Others, mainly small innovative start-ups, have developed natural raw materials. For example, Amadou Mushroom Leather, awarded the Kering Innovation prize in 2016, developed a leather made from Amadou mushrooms. Orange Fiber, meanwhile, has specialized in the creation of textiles from citrus fruit fibers. Lastly, the start-up Just Ananas makes luxury goods using Piñatex, a fabric created using fibers from pineapple leaves. Let's consider the last example to better understand the challenges of reconciling luxury and sustainable development by taking animal welfare into account.

Some luxury brands have been criticized for practices encouraging the exploitation of workers and children. What is your opinion on this issue?

"It's obviously a scandal… I am more involved with it in my work, which is plant-based natural leather. There was absolutely no regulation of anything related to leather tanning produced in Europe. Chemical products, emissions, and so on, even the harmfulness of "Crome", wasn't regulated, one of the most harmful and chemical products in the tanning industry, a chemical used in tanning to dye leather. Confronted with this, Europe outsourced production and suddenly it was countries like India and Bangladesh that had to carry out these production stages with workers who had no protection whatsoever, exposed to chemical and carcinogenic substances without knowing the risk. In some countries, such as India and Bangladesh, children, sometimes even those younger than ten, work there… when you buy a product, you don't know what's gone into making it.

"This is why it's important to talk about this and it's important that brands get engaged and carry out awareness-raising campaigns online so that people know what they're buying and so that they avoid buying products that have no manufacturing information. It's proof; when there's a lack of information on products, there has to be something going on behind the scenes."

The use of rare, animal-based raw materials contributes to the luxurious nature of luxury products. This practice is now being challenged on grounds of animal welfare. What do you think about this?

"We can have rarity without having to use animal materials. We can have rarity in cosmetics, such as floral scents that are very rare and therefore more expensive. It's the same thing in fashion and leather goods; with plants, there are products that are still little used, including the pineapple, I'm talking about that because I'm familiar with it, for now it's a niche that makes it rare and therefore prices are not low.

"It's always the same problem of communication, of the presentation of the product. If Louis Vuitton presented a range of plant-based bags tomorrow, it would have the whole Louis Vuitton brand behind it and that wouldn't change. So of course, if he launched it without fanfare by putting a bag on a corner, nothing would happen. I think that the advantage luxury brands have is *savoir-faire*, whatever the materials used.

"The brands are almost obligated to take sustainability and responsibility into account, and consumers, for their part, are more and more informed about this issue, which will lead brands to give them information. There are regular consumers who are turning away from the brands because they're better informed, there are more and more videos thanks to the development of social networks, there are loads of videos from L214. Once you've seen that type of video, it's difficult to carry on consuming in the same way. There'll never be perfect world, but if some well-informed individuals change their consumption habits, that'll be enough."

Box 2.1. *A luxury leather made from pineapple: interview with Amandine Dorin, Founder of Just Ananas*

2.3. The perceived contradiction between luxury and sustainable development: origins and solutions

2.3.1. *The sources of consumer reluctance towards sustainable luxury offers*

Including sustainable attributes in a product or service generally provides an added value to the offer and thereby makes it more attractive from the consumer point of view. Luxury presents an exception, in view of the perceived contradiction between its world and that of sustainable development. It is thus possible to identify eight sources of dissonance that help to understand the perceived gap between these two concepts (luxury and sustainable development), and thereby clarify the reasons for consumer reluctance towards sustainable luxury products. These are grouped around the dimensions of the concept, the environmental dimension and the social dimension (Table 2.2). They are illustrated below with verbatim accounts[2].

	Sustainable development	Luxury
Dimensions of the concept	Sharing Altruism Rationality Lower quality	Exclusivity Self-interest Impulsivity Superior quality
Environmental dimension	Preservation of resources	Use of rare resources, waste
Social dimension	Social equality Fairness in employment Animal welfare	Social inequality Unfairness in employment Lack of animal welfare

Table 2.2. *Sources of dissonance between luxury and sustainable development (source: (Dekhili and Achabou 2016))*

The first source of dissonance is related to the two concepts themselves, which have opposing dimensions. The first contradiction relates to exclusivity (versus sharing). If the concept of sustainable development refers to practices encouraging

2 Verbatim accounts from a qualitative study conducted among Parisian consumers of luxury items that consisted of 39 semi-directed interviews and a focus group (13 respondents). The respondents differed in age, gender and socioprofessional category. The aim was to explore the sources of dissonance between luxury and sustainable development, as well the consumer reactions to the perceived contradictions.

donations, exchanges, second-hand sales and recycling, luxury, for its part, instead evokes exclusivity, which implies the scarcity of a product:

> But this [sustainable practices] slightly harms the image of luxury products that we make, I like to think that if I buy a luxury textile product that I'm the first to have it, that it's made of high-quality materials, never touched before. I think that the sustainable aspect would put me off a little bit.

Owing to their strong emotional dimension, luxury products are little suited to the logic of donating and sharing. Consumers show an emotional attachment to the luxury products they buy (Carrigan *et al.* 2013). They associate them with cherished memories and a strong symbolism that leads them to reject the idea of having to be separated from them, even when the products reach the end of their lifespan: "if it's a question of fabric from a piece that has a history, that belonged to a star in a famous film, you might as well keep it to maintain specific symbols for a family, or an era", "I'm not sure that a woman would want to be separated from her handbag even if she doesn't use it anymore!"

The two concepts – luxury and sustainable development – are then opposed in terms of the dimension of self-interest (versus altruism). Sustainable development refers back to "others", such as the well-being of future generations, the protection of animal species and the working conditions of employees. However, luxury evokes more selfish considerations: "We talk a lot about the environment but no one pays attention, everyone just wants to enjoy themselves. I buy a product, I see whether I like it but I don't think about the consequences, I think about what I want", "especially when I want to enjoy myself, I don't think about pollution".

Another point of opposition between the two concepts relates to the dimension of impulsivity (versus rationality). When we talk about "sustainable development", we often think about thoughtful behavior that takes into account the environmental and social impact of the strategic choices and practices implemented. Luxury, perceived as being part of a care-free world, is far from rational in this regard: "Luxury: we often think of recklessness, it's true that recklessness goes well with luxury… and sustainable development is really awareness of the object and of life and of the end of the product's lifespan. There is maybe, ultimately, a difference between these two concepts." In this regard, luxury is synonymous with dreams, while environmental issues recall reality, which is not always positive.

Moreover, quality is an integral part of the definition of luxury: "For me, luxury products are a guarantee of quality. They are products that are unquestionably expensive but that, for me, symbolize a form of quality." This can be jeopardized by the inclusion of sustainable attributes. It is, in particular, the use of recyclable materials that has been subject to criticism and associated with a reduction in perceived quality:

> Recycling and luxury, that doesn't work, I'm not sure that that works… it's not in the spirit of *haute couture*, it doesn't go with luxury; the use of recycled products to make new luxury products should be avoided. That is, at a given time, they could be recycled to make other less high-end products but recycled materials shouldn't be used to make luxury products; I think that, rightly, luxury cork… some cottons, very high-quality materials, you don't get anything from recycling them.

The two concepts also seem to diverge from an environmental point of view. Sustainable development means *the preservation of the planet and natural resources*. Luxury, meanwhile, is associated with the idea of wasting resources. Indeed, the ostentatious dimension of luxury leads to a misuse of the considerable packaging: "I would say that the packaging of some products can be a bit excessive… in order to draw the eye, there is sometimes a bit much for a small item. We could maybe make reductions at this level to protect the environment." The problem of waste intensifies when rare raw materials are used (for example, leather products from exotic animals). Moreover, the specificity of luxury product categories often leads to the use of chemical ingredients that are harmful to the planet: "I imagine that when we tan hides or make fabric dyes… it can only have a negative impact. So yes, in textiles, leather goods, since there are material treatment processes that use chemicals… and we're aware that this is never very good for the environment."

Finally, at the social level, different aspects push apart the two concepts – luxury and sustainable development. Firstly, in terms of social equality, sustainable development promotes justice-related values, while luxury refers to the idea of inequalities within society. One characteristic of luxury relates to separating rich individuals from those with limited income, and dismissing people whose image is considered far from the ideal:

> There is the counterfeits issue because we see people who don't have the money and who have a luxury product, but in fact it's a fake because it gives off an image of them… so it's true that, for me, I think the biggest problem with luxury is that, it's social problems, because it's only for a specific group of people… that annoys me a bit.

Secondly, the idea of working conditions strengthens the contradiction between luxury and sustainable development. The latter advocates protecting workers and safeguarding their well-being. However, luxury often conjures negative perceptions of working conditions. The place accorded to workers in this sector remains secondary in comparison to the luxury product that garners all the attention: "The workshops that use all these small hands who make delicate things. Here's the whole chain of artisans, workers who participate in the production of the product and who are forgotten in favor of the showy aspects of the object." Moreover, workers' rights, including the case of workforces in developing countries, may be tarnished: "You know that for diamond exploration, they [luxury actors] are really exploitative." Lastly, the issue of animal welfare has provoked varied reactions from consumers; this is one of the aspects that creates a gap between the two concepts. While sustainable development "promotes the protection of animal species", the luxury sector exploits and mistreats these species to the benefit of the products created and the satisfaction of demanding customers: "A luxury product is going to be a rare product, so it's going to affect protected species. Their consumptions are going to have an impact! The use of fur, I don't know, it's kind of the first thing that comes to mind for me… because, really, it's something that affects animals that are in danger of extinction."

2.3.2. *What solutions are there for better integrating sustainable development into luxury?*

Sustainable development can be an opportunity for increasing economic prosperity and this is the challenge that luxury brands are now facing. In order to achieve this, they have several levers that can allow them, first of all, to overcome the perceived contradiction between the values they spread and those promoted by sustainable development, and then to provide an example of responsible engagement. In the following section, we will propose some potential avenues for reflection on better integrating the challenge of sustainable development into luxury.

2.3.2.1. *Building the image of luxury as a vector for sharing and of preserving resources*

Overcoming the image of companies that waste natural resources to satisfy secondary desires must be among the priorities of luxury brands. They have several arguments for this. The first concerns the essence of luxury, namely scarcity. This refers to a reasoned use of rare natural resources. The case of Ferrari, which, in 2013, decided to reduce its production by 4%, is interesting. Even if the aim behind this decision was to "encourage scarcity", the brand can also boast an objective of preserving natural resources. The second argument concerns items being passed down from generation to generation. The exceptional quality of luxury products does grant them a much longer lifespan than ordinary products. This longevity refers back to the value of sharing, as promoted by sustainable development. The watchmaker Patek Philippe is a good example of a luxury brand that capitalizes on this sharing value through its "Generations" advertising. This is an advertising campaign that was first launched in 1996 and is still ongoing. Through this message, the brand is looking to remind consumers of the exceptional quality of its products and that their longevity that makes it possible to hand them down: "You never really own a Patek Philippe. You merely look after it for the next generation."

2.3.2.2. *Sustainable luxury: a high quality that should be maintained*

The inclusion of a sustainable attribute in luxury products has been associated with a decrease in the quality level by consumers, which requires the target market to be reassured. We can highlight some initiatives to this end. For example, the Stella McCartney brand indicates on its website that the use of vegetable raw materials to replace animal-based materials has no effect on the beauty and luxurious nature of its products: "As a vegetarian brand, we never use leather, skin, fur or feathers. By taking this stance we are proving it is possible to create beautiful, luxurious products that are better for everyone – animals, people and the environment." To convince consumers that sustainable luxury products are not of a lower quality than their conventional equivalents, Stella McCartney has associated the viscose used in its luxury products (a forest-friendly material) to the excellent reputation and knowledge of its country of origin, Italy.

Furthermore, to consolidate their position as being luxury and of high quality, sustainable luxury products must be sold at the same price, or even a higher one, than conventional luxury products. Offering sustainable luxury goods at lower prices would make them more accessible, which could damage their usefulness as a symbol of ecological dedication. On the other hand, higher prices can allow consumers concerned about the status of using sustainable luxury products to display their wealth. In this regard, the example of the Lexus brand is interesting. Its decision to set the price of its environmentally-friendly LS600h model at over 100,000 US dollars was received with skepticism from experts in the sector. This positioning turned out to be valuable as sales predictions increased by 300% (Puska *et al.* 2016).

2.3.2.3. *Sustainable luxury products: a source of exclusivity and pleasure*

Exclusivity is one of the key values of luxury. It can be encouraged by high prices as well as by the scarcity of products. The use of limited editions must therefore be promoted. This is the choice made by, for example, Yves Saint Laurent for its responsible New Vintage collection. Similarly, Cadillac was able to link its altruistic values to exclusivity and personal pleasure. Through its communications on its 2016 electric ELR model, the brand indicated its engagement in favor of sustainability while insisting on the ostentatious dimension sought by luxury consumers:

> The 2016 ELR is a concretization of our engagement in favor of sustainability... The exterior of the 2016 ELR makes it a bold presence on the road: elegant, stylish and able to turn heads with ease. The minute you open the door, you notice the elegant atmosphere as well the range of functionalities that ensure safety and comfort. Conclusive proof that technology has an elegant side.

2.3.2.4. *A central place for animal welfare facilitated by technology*

Reputation is an essential strategic asset in luxury, but a particularly fragile one. In order to protect it, brands in this sector must avoid scandals related to animal welfare, which have multiplied in recent years. Beyond the above-mentioned strategies that make it possible to take this social dimension into account (such as companies purchasing their own farms for raising animals), new technologies, in particular Green Data, are another, less expensive means of vertical integration, which could enable luxury brands to ensure that their raw materials are produced in conditions respectful of animal welfare. This is the case with Stella McCartney, which has joined forces with Google Cloud to develop a traceability tool with the aim of improving its environmental and social management. In this regard, blockchain technology[3], already used by some luxury companies to protect against counterfeiting, can also be an interesting tool in improving the traceability of raw materials.

2.4. Conclusion

Sustainable development is an undeniable reality for all sectors of activity nowadays and the luxury sector is no exception. For companies, it is an issue of maintaining their legitamacy in order to prosper in a society increasingly concerned by the social and environmental consequences of human activity. This challenge is even greater for luxury companies in view of their brand notoriety and the colossal profits they create, and the fact that they are subject to intense scrutiny from sustainable development activists, in particular. In addition, considering their ability to offer innovation and creativity, luxury companies can provide an example of

---

3 A technology based on a particular form of databases distributed, secured and certified but without federal authority. A blockchain can therefore be adopted to be broadly accessible and, above all, unfalsifiable.

environmental and social engagement and bring other sectors into the movement. However, including sustainable attributes in a product does not always equal the valorization of consumers. In the case of luxury, this strategic choice could be ineffective as there can be a perceived contradiction between the values associated with luxury and those promoted by sustainable development. This chapter shows that through its tools and approaches, marketing can help to overcome this difficulty by proposing areas where the two notions can be brought together and made compatible. Luxury values as important as longevity (a long lifespan for the product, the ability of the product to be handed down from generation to generation) and scarcity (small-scale production, careful use of resources) are in line with sustainable development principles, and should be stressed. Other aspects that seem completely opposed to sustainability, such as a lack of animal welfare, must be considered. Luxury brands are expected to make a significant effort in terms of the traceability of the raw materials they use. In this regard, financial investments to ensure that animal conditions are monitored is an interesting avenue to explore. In addition, the use of technology 4.0 can make it possible to have access to complete information about animal origins.

A significant challenge facing luxury brands in the current context, which encourages environmental and social engagement, is proposing a sustainable offer while preserving the image of their products as being of an "exceptional" quality. Recycled materials, for example, are not seen as being high-quality. Environmental considerations should not be limited to recycling; the recovery of sustainable materials (for example, the fibers of pineapple and cactus leaves to replace animal leather) as a synonym for innovation, by adopting a high-price strategy, can help to maintain the positioning of luxury products. In future studies, it would therefore be interesting to explore the role of the luxury consumer in optimizing the innovation process to successfully create sustainable luxury products that are acceptable to the target audience. The literature confirms that the co-creation or even innovation by users (Von Hippel 1976) reduces the risk of product failure on the market.

Overall, luxury consumers are not the ones most concerned by environmental and social issues; their purchases continue to be driven above all by selfish motivations related to personal enjoyment, status and ostentation (Dekhili and Achabou 2016). The communication of luxury brands on sustainable development issues must therefore remain limited. However, dynamically extracted filtered information, made accessible thanks to technology 4.0 (for example, the IoT[4]), can make it possible to adapt the information communicated to the consumer profile. We

4 Internet of Things, or "connected objects": this includes all electronic devices with an embedded software system making it possible to conduct exchanges via an information and/or communication network (Internet, broadcasting, GSM, etc.).

think that a promising avenue would be to specify the type and quantity of information to mention on sustainable luxury products by taking into account the consumers' environmental awareness and need for status.

2.5. References

Achabou, M.A. and Dekhili, S. (2013). Luxury and sustainable development: Is there a match? *Journal of Business Research*, 66, 1896–1903.

Catry, B. (2007). Le luxe peut être cher, mais est-il toujours rare ? *Revue française de gestion*, 171, 49–63.

Cervellon, M.C. (2013). Conspicuous conservation: Using semiotics to understand sustainable luxury. *International Journal of Market Research*, 55(5), 695–717.

Cervellon, M.C. and Shammas, L. (2013). The value of sustainable luxury in mature markets. *Journal of Corporate Citizenship*, 52, 90–101.

Davies, I.A., Lee, Z., Ahonkhai, I. (2012). Do consumers care about ethical-luxury? *Journal of Business Ethics*, 106, 37–51.

De Barnier, V., Falcy, S., Valette-Florence, P. (2008). Comment mesurer les perceptions du luxe ? Une comparaison entre les échelles de Kapferer (1998), de Vigneron et Johnson (1999) et de Dubois *et al.* (2001). *7th International Marketing Trends Conference*, Venice, Italy, 17–19 January.

Dekhili, S. and Achabou, M.A. (2016). Luxe et développement durable : quelles sources de dissonance ? *Décisions Marketing*, 83, 97–121.

Dekhili, S., Achabou, M.A., Alharbi, F. (2019). Could sustainability improve the promotion of luxury products? *European Business Review*, 31(4), 488–511.

DeTienne, K.B. and Lewis, L.W. (2005). The pragmatic and ethical barriers to corporate social responsibility disclosure: The Nike case. *Journal of Business Ethics*, 60, 359–376.

Eisend, M. (2002). Dimensions of credibility in marketing communication. *Asia Pacific Advances in Consumer Research*, 5, 366–373.

Janssen, C., Vanhamme, J., Lindgreen, A., Lefebvre, C. (2014). The catch-22 of responsible luxury: Effects of luxury product characteristics on consumers' perception of fit with corporate social responsibility. *Journal of Business Ethics*, 119, 45–57.

Plannthin, D.K. (2016). Animal ethics and welfare in the fashion and lifestyle industries. In *Green Fashion*, Muthu, S.S., Gardetti, M. (eds). Springer, Singapore.

Puska, P., Kurki, S., Lähdesmäki, M., Siltaoja, M., Luomala, H. (2016). Male–male status signaling through favoring organic foods: Is the signaler perceived and treated as a friend or a foe? *Psychology & Marketing*, 33(10), 843–855.

Vickers, S.J. and Renan, F. (2003). The marketing of luxury goods: An exploratory study – Three conceptual dimensions. *The Marketing Review*, 3(4), 459–478.

Von Hippel, E. (1976). The dominant role of users in the scientific instrument innovation process. *Research Policy*, 5(3), 212–239.

Wagner, T., Lutz, R.J., Barton, A.W. (2009). Corporate hypocrisy: Overcoming the threat of inconsistent corporate social responsibility perceptions. *Journal of Marketing*, 73(6), 71–79

3

The Fight Against Food Waste: Approaches and Limits to Consumer-based Actions

Guillaume LE BORGNE[1], Margot DYEN[2], Géraldine CHABOUD[3] and Maxime SEBBANE[4]

[1] *IREGE, University of Savoie Mont Blanc, Annecy, France*

[2] *LEGO, University of Western Brittany, Quimper, France*

[3] *MOISA – Cirad, Montpellier, France*

[4] *Research Center, Paul Bocuse Institute, Écully, France*

3.1. Introduction

"Why should we combat food waste?" In his 2018 publication on the website of the UNESCO Chair in World Food Systems, the socioeconomist Nicolas Bricas seemed to want to provoke the reader with this title, which challenges the well-established consensus that it is necessary to combat an unacceptable level of food waste. In the text, however, the arguments put forward successfully justify the question. He explains that the issue of food waste is not a priority in terms of environmental impact compared to other resource wastage, and that global issues of food scarcity are not related to the quantity available but to distribution. Moreover, Bricas suggests "food waste is the price of freedom" and that reducing essential food waste would involve giving up certain freedoms that we are used to (the freedom to choose, not having to plan ahead, etc.). One final argument concerned the point of view to be taken: Overproduction or overconsumption? According to Bricas, "it would be better to talk about overproduction than about waste, in order to work towards a re-balancing of responsibilities and to avoid thinking that only consumers can be drivers of change, even if they can and, of course, must contribute to it."

The issue of food waste has been prominent in the media for several years, as well as in official political agendas (Europe 2020 objectives, French campaign to combat food waste) and scientific publications. As with other topics related to the environment and the overuse of limited resources, the consumer is at the forefront. How could it be otherwise?

The statistics do not appear to be in their favor, however, with households responsible for 40% of total losses and waste (outside the agricultural sector). Media, institutional and academic discussions fluctuate between blaming consumers, listing extenuating circumstances (the psychosocial biases of the consumer, the context) and calling for voluntary and simultaneous action from each of the actors in the food sector to reduce their own waste. Despite a consensus around food waste as an issue (Cloteau and Mourad 2016), the recommendations and initiatives intended to reduce waste at household and consumer level have, for the most part, limited effects.

In terms of the consumer, numerous recommendations and initiatives have appeared as part of the "enthusiasm in recent years to combat food waste". This chapter reflects on the place of the consumer in the issue of food waste. Unlike Bricas's article, we will not focus on whether reducing food waste should be a priority or not but instead will look at the scope of the recommendations and initiatives aimed at combating food waste. To do this, rather than attempting to systematically identify the limitations of the numerous initiatives that exist, we wish to spark debate around the productivist and free-market context within which the act of consumption takes place, and within which these recommendations and initiatives are at work.

In this chapter, we outline how and why some solutions for reducing waste are not effective, considering the broader dynamics in which they work. Through a movement opposed to research centering on consumers and their faults (biases), this chapter underlines the complex circumstances of food systems and the issues resulting from them, such as the level of food waste.

3.2. Food chains under tension, food losing value

Mass-market food consumption has been broadly integrated into globalized food systems. A systemic vision, one consisting of comprehending food systems in their totality and taking into account the interactions between their stakeholders, is essential in understanding the origins of food waste and all their complexity.

This view also makes it possible to highlight the contextual elements that are relevant to some initiatives intended to reduce food waste. In this section, we present some of the elements connected to the organization of food distribution systems, the priority given to the conditions under which free-trade operates, and the development of lifestyles, including the restrictions and aspirations that shape them.

3.2.1. *Food chains: the interactions and tensions of actors*

Rejected goods, price-related pressures, size requirements, precarious contracts: the relationships between mass distribution and producers, like those between agribusiness and producers, are based on power balances that are built over time and constantly reevaluated in largely tense negotiations. The same is true of negotiations between food manufacturers and distributers, as in the trade disputes that pitted Leclerc against Coca-Cola[1], Ferrero and, notably, Lactalis. At the same time, international agreements, such as CETA[2], that open up markets and put producers in global competition, work against production that responds to real (and regional) consumption needs, as well strict environmental requirements.

In this context, the consumers held "captive" by the mass market (for example, those without access to other types of retailers, or with very limited means) have little control. In the Leclerc/Coca-Cola example, consumers were faced with product shortages. With fruits and vegetables, they can either pay the price displayed, irrespective of whether it is a fair one for the producer, or do without the product. Coming back to the issue of food waste, consumers are invited to "buy according to their needs and no more". However, are we producing "according to their needs"? Planning in agricultural production would be a solution to achieve a better balance between "needs and capacities"/"supply". This can be more easily implemented in territorialized food systems than in globalized ones. Indeed, for some countries, operating in a globalized system means food production is centered on more commercial objectives (exports) rather than on securing local supply. However, "planning" goes against free trade doctrine, as shown by the existence of free trade agreements, such as CETA. Consumers, the last link in the chain, are therefore presented with an offer 1) that is not determined by real needs and 2) the price of

---

1 A dispute emerged between the E. Leclerc distribution chain and Coca-Cola following a price increase imposed by the latter. See: https://www.processalimentaire.com/flash-iaa/le-conflit-coca-cola-leclerc-s-envenime-33834?sso=1585129405.

2 The *Comprehensive Economic and Trade Agreement* (CETA) is a bilteral free-trade agreement between the European Union and Canada. Signed on October 30, 2016, it entered into force on September 21, 2017 and has been subject to some criticism.

which is the result of global competition and sector-wide negotiations. Thus, we can see the lack of consumer control, here in terms of the options available to them.

Furthermore, processors and retailers have optimization capabilities: large production volumes, consolidation of purchasing centers the option of managing stock (conservation, resale, donation), the possibility of amending prices to clear inventories, the revaluing of processing by-products, and so on. On the other hand, consumers, while numerous, are isolated and fated to act almost alone. A household can only reasonably manage a limited volume of foodstuffs owing to a limited amount of consumption, space dedicated to more than just food preservation/ consumption and so on. In addition, the members of this household do not necessarily have the option of going shopping "regularly" owing to, in particular everyday constraints on their time and space, the structure of their individual schedules and urban planning. In summary, the consumer, alone, or *isolated consumers* are not able to optimize (prices, costs, stocks, unsold/uneaten goods) in the way that agribusiness and distribution can. Initiatives such as open access refrigerators[3] – where people can put what they will not eat in time – are interesting but only concern a small number of consumers and products at any one time. These initiatives therefore only facilitate marginal improvement rather than structural optimization.

Finally, in the fight against food waste, price fixing to get rid of unsold goods is often encouraged. This is the case, for example, for "short date" promotions used by companies such as Zéro-Gâchis[4] and Too Good To Go[5] which, primarily, make it possible for retailers to avoid waste. In terms of consumer waste, the impact has not been studied. The initiatives mentioned, however, rely on mobile apps that require special attention from the user-consumer, taken from the limited supply of attention they have. Moreover, while the Too Good To Go initiative responds to the uncertain amount of fresh goods sold by artisans, it is a different matter for Zéro-Gâchis and the mass market. Is repeatedly having unsold goods a sign that supply is not matched to demand? Is it a sign that the ranges are too big or too broad to be rotated quickly enough?

In addition to the overall level of waste, it is certainly interesting to see where food is wasted throughout the production and distribution network. However, the monitoring of such waste at each level of the supply chain does not help us to

3 See: https://www.paris.fr/pages/quinze-nouveaux-frigos-solidaires-et-anti-gaspi-7018.

4 Zéro Gâchis is a start-up that helps large and medium-sized supermarkets manage their unsold goods.

5 Too Good To Go is a mobile app that allows citizens to collect unsold goods from merchants at lower cost, as there is a discount on such items.

identify the actors involved. It is a starting point for reconsidering the organization of food systems and determining, through "behavioral biases", the control that different actors have over their behavior (namely, for consumers, consumption). As we have shown, consumers find themselves alone in managing a surplus that predates their consumption.

3.2.2. *Giving value back to food?*

Another oft-proposed solution for combating waste is "giving value back to food". In fact, the hypothesis that the perceived lower value of a food product leads to a greater probability of it being wasted has been confirmed in certain contexts, such as, for example, self-service buffets (Just *et al.* 2014). While relying on good sense, this revaluing-based solution echoes a significant paradox that confronts supermarket customers. The "price war" between the different mass-market brands leads to consumer-targeted praise of the lowest prices, which, moreover, fluctuate significantly (promotions). The consumer is invited to realize that they are not paying the real price for the food. This is also true when we consider that agricultural policies allow prices to be maintained at a very low level (and a competitive one for international markets). This is even more true when we take into consideration the environmental and health implications of the production and consumption of some food products. However, the consumer develops within a universe where low prices are the norm and where promotions are also starting to become the norm (de Pechpeyrou *et al.* 2006). While it has been established that consumers predict and take into account potential waste during promotional sales (Le Borgne *et al.* 2018), repetition of such situations threatens the consistency of their final shopping cart and, more broadly, disrupts the consumer reference points for price and value.

Moreover, the increase of processed and ultra-processed products in recent decades poses a problem as it gives greater value to artificial products that may have a lesser nutritional value. The consumer is once again called on to understand the value of what they eat: this requires work from farmers, processers and, before them, nature, which slowly allowed cereals, fruits, vegetables and animals to reach maturity, be processed and brought to consumers. However, how can this be understood in the case of a product that has been ultra-processed (mainly by machines), a "black box" (Fischler 2011) made up of ingredients of multiple and not always traceable origins, and of unknown and numerous chemical additives? Can a foodstuff that has been reconstituted, retextured using reclaimed meat and extruded starch be given a symbolic value that would prevent it from being thrown away?

The demands of the market prevent a rather obvious solution – understanding the value of what we eat – from having a chance to succeed. On the one hand, this is because price-based competition interferes with the price signal and the real value of food products and, on the other hand, because the ultra-processing of products prevents food products being awarded a value other than a functional/utilitarian one. Once their ideal packaging has been stripped away, the foods lose their visual appeal and all connection with what we thought we were eating, leading to a kind of disappointment. It should be noted that, even then, the ultra-processing of food does respond to a real consumer need – having cooked meals that are simple to prepare and that have a long shelf-life – but also, and mainly, to a rationalization of costs on the part of agrifood businesses. The meat mineral market and the list of 400 additives and technological agents used by the agrifood industry are the most striking aspects of this optimized and fragmented organization that leads to the (re)constitution of ultra-processed products.

Having discussed the over-evaluation of consumer responsibility and the restrictions that limit initiatives to combat food waste, the following section highlights the importance of rules to bring about the changes expected by consumers to allow food waste to be reduced.

3.3. Consumer responsibility

The literature in sociology and marketing emphasizes that 1) throwing a product away occurs after a more or less conscious decision that the product does not meet expectations: consuming it could involve health risks; there is no longer any appetite for it after having eaten it several meals in a row; it is no longer appetizing; it is not fit to be offered to guests, and so on. Moreover, the literature indicates that 2) consumer food waste stems from a range of acts and non-acts (see, for example, Porpino *et al.* (2015)). Having a surplus of food compared to one's needs or those of one's household results in a mess of limitations and logistical considerations. It is therefore not paradoxical to state that, despite sometimes high levels of domestic food waste, most consumers hate wasting things and do not do so intentionally (Evans 2011).

The two points listed above refer, respectively, to consumption norms and lifestyles. In section 3.3.1, we look at formal and informal normative systems around food quality that have been constructed over time and encourage people to downgrade products, which leads to throwing them away *in fine*. Following this (section 3.3.2), we show how the acceleration of ways of life goes against the proposed means of combating food waste.

3.3.1. *Food standardization: An injunction to downgrade products? The case of fruit and vegetables*

It is often said that consumers have the power to change things, or, more nuanced, that they must give momentum to change. However, some key features of the products consumed are shaped by the demands of commercial actors in agribusiness and distribution.

In terms of standardization, one of the most notable examples is the selection of fruit and vegetable varieties. The major selection criteria for the varieties monitored by the French National Institute for Agronomic Research (INRA) are: appearance, regular size, high yield, long life, strong resistance to bumps during transportation and resistance to disease. Nutritional qualities and taste have been left aside. For example, the gene that extends the lifespan of industrial tomatoes for three weeks after harvesting also deprives them of their smell and gives their flesh a floury texture. The characteristics selected enable globalized and standardized trade in fresh fruits and vegetables that takes the consumer ever further from the producer. However, the "connection to the producer and the land" and the uniqueness of products are often invoked as a solution to the revaluing of products. Relocalizing the production of varieties with pleasing tastes and textures, and developing shorter supply chains could therefore be an avenue to explore. In addition to a revision of agricultural policies and of urban and rural development policies, this would require consumers to relearn how to handle and store more fragile foodstuffs carefully (and quickly).

Various initiatives have emerged in response to this standardization to encourage individuals to eat fruits and vegetables that do not conform to esthetic standards[6]. By means of such measures, manufacturers and retailers are ensuring that these products are consumed. Twisted carrots, misshapen apples and naturally non standard products now have an opening, a niche that will welcome them as long as they are distributed at a lower cost compared to their standardized counterparts. Is this really an effort to rehabilitate non-standard products? Do such initiatives give rise to new quality assessment standards? In fact, it is regrettable that these products are distributed at a lower cost, knowing that their "non-standardized" nature only very slightly affects their quality as a fruit/vegetable/foodstuff overall.

6 For example, the brand *Les Gueules cassées*, launched in 2014 by Nicolas Chabanne, that makes it possible to revalue non-standardized products using a label. The products identified as damaged are sold for 30% less than their standard equivalents. Another example is that of "ugly fruits and vegetables", started in 2014 by Intermarché, and since applied by Leclerc, Auchan, Super U and Monoprix. As with the *Gueules cassées* initiative, the "ugly fruits and vegetables" are displayed separately and have a 30% discount compared to standard products.

In comparison, damaged fruits and vegetables may not be consumed entirely or may need to be consumed within a limited timeframe, which would mean a lower price for such products. On the other hand, in the case of non-standardized fruits and vegetables, no restriction of this kind can be identified. The development of such a niche therefore only very slightly restructures the informal normative system that has developed at consumer level. To add some nuance to these statements, it should be highlighted that consuming these non-standardized products is intended to broadly reduce waste that occurs during product selection, that is, losses before commercialization. Despite everything, the message sent to consumers through the "price" signal is rather clear: these products with a "non-standardized" appearance are worth less.

3.3.2. *Combating waste at the consumer level, individualism and accelerated lifestyles: What are the contradictions?*

The question of optimizing time is now omnipresent in consumption: getting around as quickly as possible, finding information instantly, ordering and having food delivered in record time, cooking a dish that is nice but 'quickly prepared' and so on. Faced with this acceleration, the assumption is that, by optimizing daily logistics, individuals will have more time available for leisure activities and paying attention to their loved ones, themselves, their food and more. However, as Southerton (2003) points out, the acceleration of lifestyles has resulted in a widespread feeling of overwork, given the focus on convenience, which is to the detriment of attention given to care. This transition generates anxiety, coupled with an over-responsibility of consumers who, armed with logistical optimization tools, are in control of the temporal organization of their days. However, this does not translate into an increase in available time. At a more micro level, the acceleration of lifestyles results in disinvestment in individuals' actions, owing to the abundance of management tools that excessively facilitate the accomplishment of tasks (Jalas 2006). Applied to the case of food waste, we can see that consumers today have tools to facilitate their daily lives. Examples include mobile apps capable of creating shopping lists based on menus devised by the individual or by another mobile app; facilitating tools such as connected objects like smart refrigerators that track products; or packaging that indicates the edibility of the product within. Such tools embody the process (described above) of moving from "paying attention to what is around us" to "doing". In the past, it was necessary to make sure that products did not get damaged in the fridge, but the practicality now consists of information gathered on a smart phone and the product itself no longer needs to be monitored. As a result, beyond the question of how effective mobile apps are in combating food waste, it is the very idea of these tools that is challenged here: How can we promote them when they contribute to individuals feeling increasingly busy and when this busyness alters the value that these individuals place on their actions and on their food? This is therefore not a question of time that is objectively available but of a mental availability and an investment that harks back to the idea of full awareness. A more crosscutting view of the fight against food waste would therefore involve food once again being worthy of attention, in contrast to the instant management it now receives (Gjerris and Gaiani 2013).

More generally, the fight against waste at the consumer level raises the issue of individualism. Gjerris and Gaiani (2013) suggest that the tasks assigned to consumers to fight against such a large phenomenon are not commensurate with their means to redesign/modify their practices. Similarly, nowadays, the norms that are conveyed by our modes of consumption encourage individualism and hedonism (Cova and Cova 2009). However, a focus on one's own interests, or the interests of those closest to oneself, can undermine the collective view that food waste needs to be eradicated (Watson and Meah 2012).

In the following section, we discuss the need for a systemic approach in greater detail by focusing on combating food waste in mass catering.

3.4. Reducing food waste in mass catering

Mass catering is emblematic of the major challenges in our contemporary food cultures. Its changes reflect our food models as much as they contribute to defining them. While the imperatives of public health, hygiene and nutrition shaped, and continue to shape the sector, the 21st century has come with new challenges in limiting the impact our food has on the environment, including limiting food waste. Faced with the need for rapid action to achieve the reduction objectives set by public authorities[7], the initiatives intended to combat food waste in canteens often emphasize consumer awareness in order to limit leftovers. A key approach in this regard consists of collecting, separating and then weighing the food waste at different stages of the catering process – storage, processing, distribution, consumption – before reporting the results to the guests by, very often, combining the weight of the waste with understandable equivalents in terms of monetary or environmental impact. The consumer thus "informed" of the scale and consequences

7 In France, the adoption of an initial national pledge to combat food waste in 2013, followed by a second pledge in 2017, as well as the adoption of legislative provisions (Article 102 of Law No. 2015-992 of April 17, 2015 on energy transition for green growth and Law No. 2016-138 of February 11, 2016 on combating food waste), have made it possible to start a national process towards achieving a reduction target of 50% by 2025.

of the food waste is then supposed to change their attitude to the issue, make changes voluntarily and become aware of their choices and consumption in order to "clear their plate".

Two questions arise from this: To what extent can individual efforts contribute significantly to reduction objectives? To what extent should consumer attitudes towards food waste be the priority action to combat waste in mass catering?

3.4.1. *Separate, weigh, and inform: A winning strategy?*

The first question invites us to discuss the extent of consumer responsibility. Owing to strong methodological constraints (human, organizational and material), the quantification of food waste is carried out almost systematically at the collective level. The food left on most plates is collected, potentially separated by the type of food (for example, vegetables, meat) or type of waste (for example, avoidable, unavoidable) and then weighed as a whole to gauge the total weight of the waste accumulated at the catering site. This weight is then expressed in average grams per person (that is, the total weight divided by the number of meals served). This approach has the benefit of providing an overview of the overall situation while limiting the restrictions related to the quantification of food waste. Nonetheless, the aggregation of individual data paints an impressive picture of the scale of food waste and gives the impression that all consumers waste, and that they do so in equal measure. However, the work carried out to evaluate the inter-individual variation in waste shows a significant heterogeneity in the amounts observed and emphasizes that a majority of individuals tend to leave either nothing or very little on their plates. For example, the results of eight days of weighing in the restaurant of a research and higher education center in the Montpellier region show that, from 1,146 meals, 25% of the plates had no food waste at the end of the meal, while half of the plates had less than 15 grams of consumable food left (Sebbane 2018). Other research also shows that a large proportion of adults waste little to nothing when eating in public (Al-Domi *et al.* 2011). Lorenz *et al.* (2017) note, from a visual assessment of food waste in a German company restaurant, that 75% of diners leave almost no food at the end of a meal.

As a result, when we bring this down to the individual level, it appears that efforts to reduce food waste do not affect the majority of the population and that the priority must be to find out the reasons why some guests, at a given time, waste much more than others. However, the factors with the greatest impact on waste reveal less a lack of consumer awareness and more parameters related to the organization and offering of catering, in which individuals have little room to maneuver. Indeed, while wasting is largely considered to be a socially and morally undesirable act, contextual factors, such as the failure to adapt the portions served, or the perceived low quality of taste of the meals, act as barriers that prevent or hinder the translation of good intentions to reduce food waste into effective action (Sebbane 2018).

Consequently, affecting the motivations of individuals without giving them the means to assume their responsibility runs the risk of, at best, creating acceptance of the prevention message without any lasting changes, and, at worst, causing skepticism. Generalizing the introduction of different portion sizes for all components of the meal, or offering the possibility of "tasting" the dishes by developing test tables upstream of self-service restaurants are examples of actions that make it possible to reduce certain barriers, by giving consumers more control over adapting their choices to meet their desires.

As a result, while national campaigns to combat waste and multiple awareness-raising activities have largely contributed to making the issue part of the collective consciousness, it is now necessary to take into account consumers' actual degree of autonomy. The behavioral objective inherent in social marketing reminds us that gaining acceptance of an idea (for example, that we have to combat food waste) or changing a belief (for example, food waste has an impact on the environment) cannot be considered a successful intervention if it does not translate into actual change in individual and collective behaviors (Rundle-Thiele 2015).

3.4.2. *Towards a collective awareness of sectoral restrictions and the degree of consumer autonomy*

While consumers face up to the restrictions that limit their fields of action, it is also necessary to highlight the economic, hygienic and nutritional imperatives that the sector's managers and staff must respond to in order to address their more or less direct impacts on food waste. The social and non-profit vocation of catering services is embodied in a highly subsidized meal price and, consequently, a cost for the entities in charge of service. The managers and decision-makers are thus subject to a constant search to optimize the resources used (human, technical and logistic), implying decisions that negatively affect the quality of foodstuffs. This budgetary pressure also leads to the development of central kitchens and cold preparations, which are economies of scale but also industrialization of the preparation process. Throughout the 20th century, up until today, hygiene and traceability regulations have been strengthened in accordance with technological and organizational innovations, as well as in response to consumer concerns about a food system regularly marred by scandal. The legislative framework clearly places the reduction of risks of foodborne illnesses above aims to reduce waste (as illustrated by the strict framework for the reuse of excess food). This hygiene pressure has also contributed to promoting the use of semi-finished food products from agribusinesses, to transfer some of the health risk to actors upstream of the catering system.

At the same time, the cooking profession has itself been profoundly changed by these imperatives. Responsible for the bacteriological quality of the meals, chefs have to comply with numerous procedures (HACCP, traceability, health control plans) that affect the time dedicated to enhancing the taste of the meals. The complex organization of mass catering, which stems from a set of legislative, social and economic constraints, is therefore not without consequences for the perceived quality of the dishes offered and contributes to an increase in the affective distance between the guests and their meals.

In taking mass catering as an example of a food sub-system, this section shows how the issue of reducing food waste goes beyond mere "information" on the scale and consequences of the problem. We emphasize that the consumer's degree of autonomy and the structural constraints imposed on the sector must be jointly considered, in order to determine relevant and lasting actions that can be taken. Rather than implementing one or several actions considered and developed "from the outside", social marketing encourages collaborative approaches. Developing areas of consensus to co-create a diagnosis that goes beyond the search for responsibility, and leads to the definition of universally accepted actions is a path to be encouraged. Box 3.1 provides an interview with a professionnal involved in a project to reduce food waste in mass catering. It highlights that the relative consensus about combating food waste is a unifying asset that can stimulate collective awareness among users, catering agents, managers and decision-makers, and that this makes it possible to begin joint and broad considerations of the future of our food system.

Luc Lignon, Director of Food Policy for the City and Metropolitan Area of Montpellier shows us a holistic approach through the project "Ma cantine autrement" *("Doing my canteen differently"). This project is aimed at promoting sustainable food in the 89 school canteens in the city of Montpellier, while satisfying requirements related to the self-managed production and distribution of 14,600 meals per day in a central municipal kitchen.*

Reducing food waste through the "Doing my canteen differently" program

The approach developed in elementary schools in the City of Montpelier is part of the assessment that food waste is not so much a matter of specific individual behaviors, but the result of multiple factors throughout the catering system. Lignon explains that "food waste is not related to people specifically but more to a system. Food waste must be controlled at all levels of the process; this starts with buying and continues throughout the process, up to the distribution of meals." Acting on food waste requires questioning all of the stages: "When I buy something, what is the quality of the raw material, how is it received, stored, cooked, processed, refrozen and made appealing when it's given to children". In addition to wastage that can occur at various stages, these stages impact the volume of waste that follows: "All actions have an impact; if I present an apple well thanks to the cutting kit set up for the activity leaders, and the apple is bad, the child still won't eat it." For these two reasons, it is not enough to focus on making children aware that they should clear their plates.

"Doing my canteen differently" is a systemic program for managing school catering and was established during the period 2015-2020 with the aim of creating a sustainable food system. It brings together 18 activities designed around four interdependent areas for improvement, developed simultaneously: the redesign of food purchasing policies (for example, the structuring of local supply networks, changing tenders and allocations); the improvement of product management (for example, a new ordering system, adaptation of food plans, recipes and weights in accordance with the continued measurement of satisfaction); the modernization of meal distribution (for example, the establishment of self-service facilities, staff training, fruit cutting kits); and raising awareness of sustainable food (for example, sorting tables, eco-citizen meals, awareness-raising events, educational leaflets).

Impacts of the systemic approach to food waste

In terms of the effectiveness of the "Doing my canteen differently" program, Lignon suggests that "the reduction of waste is not an end in itself", before emphasizing that a target of zero waste cannot be achieved without compromising on the hygiene, nutritional and social inclusion requirements that are at the heart of mass catering. He states that it is fanciful to believe that the service and meals suggested will respond to the large range of child profiles, despite optimization of the production and distribution process: "There will always be waste." As a result, for the Director of Food Policy, "the fight against food waste is more a means of creating a dynamic around issues of sustainability and allowing the interaction of levers of action", which goes beyond the mere issue of reducing food waste. Nonetheless, he highlights the impressive results on food waste: "Our comprehensive approach has produced encouraging results with a reduction of around 27% in food waste – throughout the system – within four years, and this has made it possible to save on purchases of around 70 tons of foodstuffs per year."

Box 3.1. *Interview with Luc Lignon about the "Doing my canteen differently" project*

3.5. Conclusion

This chapter is aimed at putting our understanding of food waste into perspective, not only by revisiting the most recent conception of the issue, which makes the consumer the key actor in driving action, but also by questioning the feasibility and the lack of systemic vision of the actions already implemented.

We have shown how a trickle-down approach to responsibility occurs throughout the food chain to reach the consumer. Power relationships are developed, guiding the outcomes of tough negotiations between producers, manufacturers and retailers. Despite this, any questioning upstream of agricultural sectors does not take place in terms of production planning but rather takes the form of an offer being put on the market that does not necessarily fulfil needs. This offer – produced worldwide – is mainly guided by imperatives of profitability, with production and processing choices fluctuating between the prices of agricultural goods, production and processing costs, and the market price. Strong competition in the mass-market sector and the race for low prices (which results in widespread price decisions and thus the above-mentioned transport-related requirements) poses the question of the value allocated to products through a "price" signal. In a more crosscutting manner, at the level of society, the chapter reports on how individualism, which is ever more important, reduces the likelihood of individuals acting in the common interest, and that the anxiety-inducing situation of overwork that people are experiencing (in particular, owing to the proliferation of digital devices) leads to less time and care being spent on food. This brings with it a loss of connection with food. With this in mind, what are the conclusions on the use of social marketing to support sustainable development?

Firstly, our remarks do not refute the idea that reducing food waste is necessary and that this occurs through, among other things, changes to individual behaviors. However, in order to achieve this, the role of the consumer must be put in context and the issue of food waste must be part of a program of societal change, in order to move towards production, processing and distribution systems that make it possible to "produce and consume less but better".

The search for "sobriety" (not buying too much, not heating up too much, keeping clothes for a long time, taking the right steps to avoid wasting food, etc.) is an increasingly widespread trend in society. However, its spread is running up against different conceptions of consumption. The act of consuming can be a means of both belonging to a group and adhering to certain values. Frugality can thus be associated with responsible, socially conscious consumption or, conversely, be perceived as a sign of limited purchasing power. The idea of the "smart shopper", disseminated by environmental marketing, instead highlights the ability of the individual to get the best deals (buy more at lower cost) in order to serve their own interests, rather than their ability to buy in moderation without succumbing to temptation, thereby contributing to the common good. In this regard, the marketing strategies used (widely) and the cognitive, emotional and behavioral impulses which they rely on, promote excessive buying, consumption (Chandon and Wansink 2010) and waste, for which consumers are blamed. Winning the battle for frugality that would be widely understood as a virtue, and not as a class marker, cannot be done through isolated communication campaigns (the media, billboards, events, etc.) that would settle for addressing the right message to the right target audience at the right time. The complexity of issues related to food waste (and the environmental transition more broadly) involves changes at the micro, meso and macro levels which would involve extending the traditional field of research of social marketing centered on consumer behavior. Joining marketing research to other disciplines, such as economic, political science and strategic and institutional approaches in management sciences would make it possible to break down the barriers between different types of knowledge. It would also provide more comprehensive analyses in order to fully consider the complex structures that determine behavior and structure the real capacity for consumer behaviors. In doing so, this approach would help public authorities articulate the necessary structural, legislative and educational measures to guide societal change (Luca *et al.* 2016).

Promoting a bottom-up approach by involving different communities of actors in shared and co-constructed initiatives also appears to be a promising area in this field. The traditional application of targeted tools for segmentation and the "marketing mix" have actually been the subject of criticism. This is, on the one hand, because of the dominant role they give organizations in determining value, and, on the other hand, because of the unidirectional transmission of these messages, from social marketing specialists to specific groups (Luca *et al.* 2016). However, the challenges posed by more sustainable development are accompanied by a necessary shared understanding of the problems and solutions between multiple groups of actors, which requires recognition of the active role of different parties in the creation of values. Initiatives such as the consumer brand "Who's the boss?!" ("*C'est qui le patron*?!")[8], illustrate this approach by discussing consumer needs and expectations to develop responsible food products that take into account the rights, responsibilities and capabilities of all actors in the food system network.

8 "Who's the boss?!" ("*C'est qui le patron*?!") is a brand that was created in 2016 with the aim of enabling consumers to create the specifications of a product (manufacturing and commercialization). Between 2016 and 2020, 30 products have been created. See: https://lamarqueduconsommateur.com/.

3.6. References

Al-Domi, H., Al-Rawajfeh, H., Aboyousif, F., Yaghi, S., Mashal, R., Fakhoury, J. (2011). Determining and addressing food plate waste in a group of students at the University of Jordan. *Pakistan Journal of Nutrition*, 10(9), 871–878.

Chandon, P. and Wansink, B. (2010). Is food marketing making us fat? A multi-disciplinary review. *Foundations and Trends in Marketing*, 5(3), 113–196.

Cloteau, A. and Mourad, M. (2016). Action publique et fabrique du consensus. *Gouvernement et action publique*, 1(1), 63–90.

Cova, B. and Cova, V. (2009). Les figures du nouveau consommateur : une genèse de la gouvernementalité du consommateur. *Recherche et applications en marketing*, 24(3), 81–100.

Evans, D. (2011). Blaming the consumer – Once again: The social and material contexts of everyday food waste practices in some English households. *Critical Public Health*, 21(4), 429–40.

Fischler, C. (2011). L'alimentation, une consommation pas comme les autres. *Les Grands Dossiers des sciences humaines*, 3, 9–9.

Gjerris, M. and Gaiani, S. (2013). Household food waste in Nordic countries: Estimations and ethical implications. *Etikk i praksis-Nordic Journal of Applied Ethics*, 1, 6–23.

Jalas, M. (2006). Making time: The art of loving wooden boats. *Time & Society*, 15(2/3), 343–363.

Just, D.R., Sığırc, Ö., Wansink, B. (2014). Lower buffet prices lead to less taste satisfaction. *Journal of Sensory Studies*, 29(5), 362–70.

Le Borgne, G., Sirieix, L., Costa, S. (2018). Perceived risk of food waste: Influence on consumer attitudes towards and choice of sales promotions. *Journal of Retailing and Consumer Services*, 42.

Lorenz, B.A., Hartmann, M., Langen, N. (2017). What makes people leave their food? The interaction of personal and situational factors leading to plate leftovers in canteens. *Appetite*, 116, 45–56.

Luca, N.R., Hibbert, S., McDonald, R. (2016). Towards a service-dominant approach to social marketing. *Marketing Theory*, 16(2), 194–218.

de Pechpeyrou, P., Parguel, B., Mimouni, A., Desmet, P. (2006). Valeur et sincérité perçues d'une promotion multi-mécanismes. *Recherche et applications en marketing*, 21(4), 25–39.

Porpino, G., Parente, J., Wansink, B. (2015). Food waste paradox: Antecedents of food disposal in low income households. *International Journal of Consumer Studies*, 39(6), 619–629.

Rundle-Thiele, S. (2015). Rétrospective et perspective d'avenir : un plan d'action pour la recherche en marketing social. *Recherche et applications en marketing*, 30(3), 133–139.

Sebbane, M. (2018). Réduire le gaspillage alimentaire en restauration collective : analyse des déterminants psychosociaux et évaluation d'une intervention comportementale. PhD thesis, University of Montpellier, France.

Southerton, D. (2003). 'Squeezing time': Allocating practices, coordinating networks and scheduling society. *Time & Society*, 12(1), 5–25.

Watson, M. and Meah, A. (2012). Food, waste and safety: Negotiating conflicting social anxieties into the practices of domestic provisioning. *The Sociological Review*, 60(S2), 102–120.

4

Food Waste in Family Settings: What are the Challenges, Practices and Potential Solutions?

Amélie CLAUZEL[1], Nathalie GUICHARD[2] and Caroline RICHÉ[2]

[1] *Paris 1 Panthéon-Sorbonne University, ISC Paris, France*

[2] *RITM, Paris-Saclay University, Sceaux, France*

4.1. Introduction

Devoting a chapter to family food waste can be justified by two things: on the one hand, the enormous challenge that the fight against food waste presents and, on the other, the role of the end consumer in this fight within their own home .

Defined by the deterioration, removal or throwing away of still-edible food, waste can stem from either deliberate or non-deliberate actions (Leborgne *et al.* 2018). A truly global problem, it represents a a huge loss of food across the world (1.3 billion tons per year worldwide[1], 10 million tons of which are in France[2], or 29 kg per person in France[3]), a considerable economic cost, an ethical scandal (if we consider that a quarter of the waste would be enough to feed one eighth of the global under-nourished

1 http://www.fao.org/3/i2697f/i2697f.pdf.

2 Ministry for Environmental Transition, April 2018.

3 http://www.ademe.fr/etat-lieux-masses-gaspillages-alimentaires-gestion-differentes-etapes-chaine-alimentaire.

Marketing for Sustainable Development,
coordinated by Sihem DEKHILI.

population), a squandering of resources and, indeed, is a significant contributor to global warming (in France, "the lifecycle of the average French person's food represents around 20% of their total daily emissions"[4] (Masson and Gojard 2019)).

Food waste is a major challenge at an economic, social, ethical and environmental level. If this waste is attributable to different causes and produced at all stages of the food production and consumption process, the family, at the end of the chain, is undeniably one of its actors. In fact, at the European level, consumers are responsible for 53% of products that are thrown away. Each year 20% of all food waste produced in the European Union is wasted, which is some 88 million tons of food. While the dunce's cap goes to the Netherlands (with 541 kg/inhabitant) and first place goes to Slovenia (72 kg/inhabitant), France is in 16th place in the ranking of European countries by food waste (136 kg/inhabitant) with significant room for improvement[5].

In France, the national pact to combat food waste, signed in 2013, is based around 11 measures and is aimed at reducing food waste by 50% by 2025. This means an average annual reduction of 5% per year between 2013 and 2025.

From this perspective, all actors in the production chain are involved and their fight has already begun, particularly since the Energy Transition for Green Growth Act of 2016. This requires the establishment of an approach to combat food waste. Citizens, beset by a new "responsible conscience" (Nabec and Roux 2016) are also mobilizing more and more, particularly in the context of the choices they make at home. Despite the methodological difficulties in accurately measuring intrafamilial waste (owing to the intimacy of the context, the variety of possible measurements and even the definition of waste itself), the figures remain telling and demand that interest be taken in the phenomenon of waste in family homes.

However, family decision-making dynamics have been little studied in terms of waste and the production of food waste (Stefan *et al.* 2013). Some works (for example, Schill and Fosse-Gomez (2014)) nonetheless seek to understand waste in a family context, not only from an individual perspective. Studying the more or less responsible practices within families and their decision-making dynamics also requires consideration of the agents of socialization that have an impact on the family unit, such as schools (and canteens)[6].

---

4 See: www.rodez-agglo.fr, available at: https://www.rodezagglo.fr/uploads/2019/05/carnet-de-la-prevention-restaurateur.pdf.

5 https://www.touteleurope.eu/actualite/88-millions-de-tonnes-de-nourriture-jetees-chaque-annee-en-europe.html.

6 But also when the family eats outside their home (in a restaurant for leisure or work). This aspect of family food as a possible source of waste is not addressed in this chapter.

This chapter will thus present the roles of family members in relation to food waste and will then address the multifaceted aspects of waste, in line with the stages of the family consumption process (managing shopping, putting it away, eating, being selective or even throwing things away during the course of a meal). A range of practices that are already in place or that could be of interest to better manage this phenomenon will also be proposed.

4.2. The actors in family food waste: everyone is involved!

Owing to varied places of consumption and multiple actors, food waste is multifaceted and its sources are numerous. Within families, it is natural to link it to the food consumption process of the family members and the management of "food duty". It is also explained by the varied consumption and purchasing situations of families, which are characterized by a large variety of sociodemographic, cultural and educational profiles that have an impact on their relationship to food and waste.

This diversity (of actors and situations) makes the precise quantification of food waste within family groups a delicate proposition. Indeed, members of the same family may not have all their meals at home (some may be in a school, workplace or restaurant, for example). Furthermore, the very notion of a meal varies from one family to another (for example, meals bringing all family members together versus staggered snacking dependent on when each member returns home).

4.2.1. *One family, one way of wasting: many families, many ways?*

Talking about food waste within families suggests that the family is considered as a homogenous entity. However, on both the sociodemographic and internal functioning levels, the family is rather reminiscent of a plurality of groups, and as such can potentially generate varied waste-related situations and sources of waste.

4.2.1.1. *Features of families and waste*

First of all, the family can be considered in terms of its numbers, as well as its "qualitative composition" or its habits. Regarding its *quantitative aspect* (the number of people composing it), the family brings to mind very different realities in terms of *a priori* risk of waste, which has consequences on a number of levels:

– quantity of food to provide: it is easier to work out the quantities that will be consumed during a meal when there are fewer people; on the other hand, a higher number inevitably leads to approximation, a potential source of waste;

– type of food to provide: the higher the number of people making up a family, the more a planned menu will be subject to changes that may or may not be predictable. Indeed, the differing tastes of members, their potential intolerances and allergies, contingencies relating to guests and, more generally, the behavioral norms that characterize the family can create disparities between the menu that was planned and the dishes that will end up being eaten.

The *qualitative composition* of the family also plays a role in its propensity to waste or lack thereof, as well as the perceived responsibility of each member. Without claiming to be exhaustive, some factors related to the characteristics of the family (Palatnik *et al.* 2014) and its members seem to naturally encourage or limit food waste (Table 4.1).

<table>
<tr><th></th><th>Characteristics</th><th>Possible consequences and effects in terms of food waste</th></tr>
<tr><td rowspan="5">Sociodemographic and sociocultural factors and level of food waste</td><td>Family income</td><td>Higher or lower, making it “necessary” to varying degrees to reuse potential food waste from meals, as well as being careful with the products bought and used to limit waste</td></tr>
<tr><td>Profession of adults</td><td>More or less demanding and leaving dedicated time to manage shopping, the fridge/cupboards and varied meals</td></tr>
<tr><td>Presence of children</td><td>Varying numbers of siblings, with different ages and rhythms (potentially making it complicated to plan meals at home), increasing the risk of waste</td></tr>
<tr><td>Family rules or habits, behavioral or educational norms</td><td>Meals taken together or not, the possibility of each member choosing the composition of their own meal or not, leaving what they do not like, not attending family dinners, warning of absence from a meal in advance, adapting to individual tastes or not, etc.</td></tr>
<tr><td>Type of family (more or less traditional), distribution of roles within a couple or family</td><td>Fixed or fluctuating: a shared responsibility for meals and shopping, implying deliberate organization; on the other hand, sole responsibility for food and waste resting with a single family member</td></tr>
</table>

	Characteristics	Possible consequences and effects in terms of food waste
Sociodemographic and sociocultural factors and level of food waste	Familial organization	Planned menus or, conversely, leaving room for spontaneity or unpredictability – unexpected meals out, surprise invites, etc.
	Awareness of sustainable development and the environment[7]	Higher or lower and shared, making the issue of waste more or less central
	Variety of tastes	Makes it complicated to create a single menu and potentially increases the risk of leftovers that may or may not be used
Heterogeneity of members and associated risk of food waste	Autonomy of members	Varies according to their age and status (adults versus children) and increases the difficulty of managing food "well"
	Presence of animals	Can, depending on the type, create additional work in managing food. On the other hand, the presence of a dog, for example, can prove useful in the reuse of (some) food waste
	Family member who is particularly mindful of waste (strong environmental concerns) and is listened to within the family	Can, through their presence alone (dissemination of knowledge, pro-environmental behaviors), lead their previously uninterested family towards more responsible behavior

Table 4.1. *Impact of family characteristics on the generation of food waste*

4.2.1.2. *Families with and without children: what are the differences?*

Attempting to answer this question first leads to consideration of the waste of families with and without children, as the number of children inevitably increases the sources and risk of waste. Over 70% of young parents confirm that the arrival of

7 Sustainable development is a necessity for 71.9% of French people, while 60.7% are engaged through their consumption and way of life (Greenflex 2016). Available at: https://www.lesechos.fr/2016/09/de-plus-en-plus-de-francais-consomment-durable-214

a child changed their food habits[8]. They cook more, they make more use of fresh produce, the price is no longer systematically the first determinant in choices and organic produce is chosen more often. However, does this arrival also mobilize the parents in terms of waste? It is difficult to give an answer to this question as the parameters to consider are varied. However, if the child is completely dependent on the food choices of their family, especially when very young, when growing up they will progressively confirm their personality and their tastes, which are considered to a greater or lesser extent depending on the type of family. For instance, many children turn their noses up at a plate of cauliflower or spinach. Some parents will insist that their child tastes it, or even that they clear their plate, while others will pay less attention to the fate of the contents of the plate, with a risk of waste into the bargain (Box 4.1). While the presence of children in a family naturally increases the sources and risks of waste, it poses three questions about food waste that will be discussed in section 4.2.2:

– Why should and how can children be taught not to waste?

– What role can the child play in controlling family food waste?

– Which actors provide the child with knowledge about waste?

The food-related norms and practices passed on to children vary between families and reflect different food-related educational models (Lalanne and Tivert 2008):

– in Emma's and Kevin's family, there is no room left for changes to be made: the menu is set and no alternative is proposed. Emma is required to clear her plate; Kevin is allowed not to finish his meal but there is no alternative and it's tough luck if he's still hungry;

– in Paul's family, changes can be made: his parents follow an "informative and preventative" approach and want to provide their child with information that will allow him to make food-related decisions in a responsible manner.

Educating children confronts their parents with the need to make decisions between potentially contradictory situations: introduce their children to different tastes, make the children eat when they aren't hungry, stop them when they've had too much, promote health over pleasure and to try and balance the two, limit food waste…

Box 4.1. *The food-related education of children, or when health, pleasure and waste become potentially contradictory*

8 http://www.leparisien.fr/week-end/alimentation-bebe-les-parents-en-parlent-10-03-2017-6744912.php.

Childhood is a crucial stage in instilling good habits, particularly those related to food (healthy, balanced and varied eating), and responsible behavior (Box 4.2). Indeed, it is preferable, as it is more efficient and long-lasting, to help individuals to develop good habits in childhood (for example, desirable attitudes and behaviors in regard to waste), rather than trying to alter habits once they have been established (Freeman and Brucks 2001, pp. 243–252; Charry *et al.* 2012).

The fight against food waste within families can, paradoxically, lead to children being overfed, particularly in underprivileged families (Kral and Hetherington 2015). The aim must therefore be to promote improving the balance and nutritional quality of the food of young consumers, while also reducing waste.

For National "*Anti-Gaspi*" ("against waste") day in France (October 16, 2019), the Ministry of Agriculture and Food renewed its PR campaign on "stories from our childhood" in which popular heroes provide examples of the fight against waste:

– "Mr Crow no longer drops his cheese";

– "Hop-o'-My-Thumb understands how precious bread is";

– "at midnight Cinderella's pumpkin will make a great soup";

– "when Donkey Skin bakes her cake, the prince doesn't lose a crumb";

– "nothing is lost and everything is frozen with the Snow Queen";

– "Snow White should have munched on an ugly apple";

– "the Three Bears' porridge isn't left for Goldilocks".

As it was popular among both children and adults, this campaign spoke to the public at large and strengthened intergenerational links on the issue of combating waste.

Box 4.2. *When the Ministry of Agriculture and Food took action against food waste*[9]

4.2.2. *Role and perception of the main members of the family on food waste*

Each member of the family has a role to play in relation to waste but the question is not posed in terms of who is an adult, child or teenager.

9 https://agriculture.gouv.fr/les-contes-de-lantigaspi-le-ministere-se-mobilise-contre-le-gaspillage-alimentaire.

4.2.2.1. *Adults and waste*

Whether or not they are parents, adult consumers have a heterogeneous view of what constitutes a "responsible consumer" and differing responses. According to an Ethnicity/ADEME[10] study in 2017, 52.9 % of French people try to be a "responsible consumer" by buying in a "less superfluous" manner or doing so "differently" (products that are labeled ethically certified, local, less polluting, etc.), without necessarily actually reducing their consumption. While there is still a noticeable gap between the intention to consume better and the reality, the phenomenon of deconsumption is still effective and it intensified in 2019[11]. However, customers are not deserting stores; they are buying, in particular, more organic or regional products, which often have higher prices. When they are in a couple, the adults in a family often mention food waste. This issue can be a source of conflict, but can also lead to compromises and, in the longer term, to a kind of mutual learning (Clauzel *et al.* 2016).

4.2.2.2. *The child-consumer, socialization and waste*

A child learns to become a consumer through a *process of socialization* that takes place by means of different socialization agents, including, of course, family and school, but also peers and the media. During childhood, the role of agents of socialization is thus crucial in learning about practices and norms, in particular those related to sustainable development and responsible consumption.

The fact that a family has children leads to the question of intergenerational influences: *who influences whom and to what extent?* (Gollety *et al.* 2012, pp. 223–254). If the child becomes a consumer by imitating those around them, then a phenomenon of inverse socialization can also occur within the family (Gollety *et al.* 2012, pp. 223–254). In a suitable environment and under certain conditions (the child understanding certain concepts, schooling, encouraging children to express themselves), the child becomes capable of teaching their parents. This inverse socialization has been confirmed in the fields of food and food waste. As a result, new eco-citizen and anti-waste actions would be passed on more and more by children to their parents and not only in line with a traditional top-down approach (Clauzel *et al.* 2016). Three types of responses can be identified (Figure 4.1) when food waste at home is discussed with children.

School-age children will be inclined to protect the environment sooner, even if they would not often initiative responsible actions (Box 4.3) at home, such as sorting waste (Schill and Fosse-Gomez 2014).

---

10 French Environment and Energy Management Agency.

11 https://www.lesechos.fr/industrie-services/conso-distribution/la-deconsommation-installe-pour-les-produits-du-quotidien-1128301.

Supervising food waste

- The child must clear their plate.
- The amounts they consume are checked by their parents.
- Waste is not mentioned much by their parents but it is expected.

Explaining food waste

- The child receives explanations related to the impact of food waste (reasons why they must not waste).
- The child tends to consider the issue of waste.
- The child asks more often to be served more reasonable and considered amounts.

Ignoring food waste

- Waste is rarely addressed in the family.
- Annoyed by guilt-inducing discussons, the parents see checking or monitoring food waste (for them and for their children) as a limitation.
- The child has no restrictions on the amount they eat or throw away.

Figure 4.1. *Three ways of addressing food waste at home with children (source: Clauzel* et al. *2016)*

Why not establish a soup game for everyone in the family? Every weekend, suggest to the children that you all make a soup with the leftovers in the fridge and everyone in the family takes a turn to choose a mystery ingredient that the others have to identify when they taste it. This solves two problems: that of getting children to eat vegetables and that of managing the leftovers in the fridge, all while educating the younger generation in an entertaining manner.

Inspired by (Desbordes and Botavant 2005).

Box 4.3. *Soup for everyone! Or how to instill good anti-waste instincts in children*

4.2.2.3. *Teenagers and waste: perceptions and responses*

Adolescence is a period ranging from around the ages of 12 to 18 that involves changes, increasing independence and an affirmation of identity (Ekström 1995). It implies a kind of maturity and autonomy when it comes to choosing a product (Ekström 1995). Adolescence also shows a certain vulnerability, owing to a lack of experience (Pechmann *et al.* 2005). As a result, it appears to be a key stage in acquiring consumer skills. In the same way as children, teenagers are largely guided by their parents and schools (Box 4.4) in terms of food education (Nicolas-Hémar *et al.* 2013), as well as in terms of environmental awareness and are strongly influenced by the standards of their families and the actions of their parents (Grønhøj and Thøgersen 2012).

> Max-Jacob School in the Loiret region has managed to reduce the amount of food thrown away by pupils in the canteen by a third by implementing a number of anti-waste initiatives (a maximum of three pieces of bread per person, delicately carefully presented plates, etc.), and, in particular, a proposed swapping table. "In concrete terms, a pupil goes through the self-service line and takes cheese and dessert, for example, before realizing at the end of the meal that they don't want their cheese. They can then put it on a dedicated table. Another pupil who is still hungry can then eat the leftovers. Without this table, the food would end up in the trash."

Box 4.4. *A school in the Loiret region sets up a "swap table"*[12]

While the teenagers have taken on board the concept of food waste well (Clauzel *et al.* 2017), for them it remains subjective and very dependent on context. The act of wasting is accepted but the responsibility for this action is often deferred to other actors (parents, schools), depending on the context (Figure 4.2).

The cause of waste is thus attributed to others (family, school), while, on the other hand, when these young consumers do not waste or waste little, they tend to attribute the benefits of this low level of waste to themselves (Clauzel *et al.* 2020). This analysis reflects a causal attribution process (Kelley 1973) according to which individuals, here teenagers, attribute good behaviors to themselves (self-attribution) and give the less rewarding behaviors to others (hetero-induction).

12 https://www.lefigaro.fr/conso/2018/10/16/20010-20181016ARTFIG00006-contre-le-fleau-du-gaspillage-alimentaire-un-college-installe-une-table-de-troc.php.

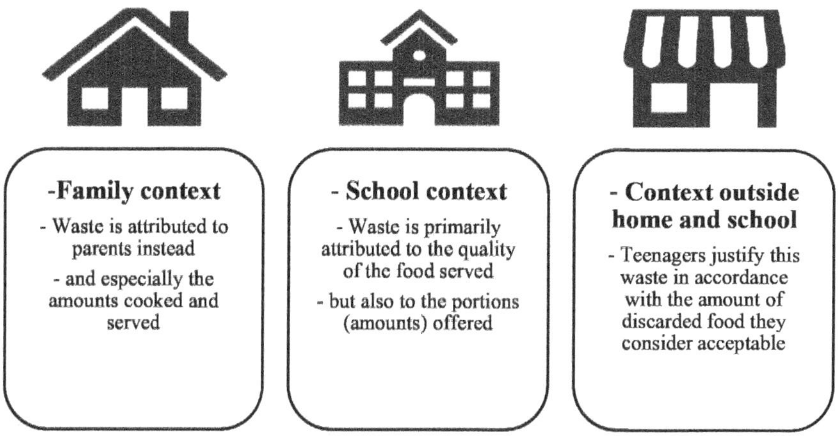

Figure 4.2. *Three ways of explaining food waste by teenagers (source: Clauzel et al. 2017). For a color version of this figure, see www.iste.co.uk/dekhili/sustainable.zip*

4.3. Multifaceted wastage during family consumption at home

Changes to family functioning, the role and place of different family members in the numerous tasks associated with organizing meals, food shopping and its storage are some of the important factors in waste. Who is the decision-maker/in charge of this task? Who does what? Who feels responsible for what?

The organization and distribution of household tasks has undoubtedly evolved in recent decades owing to the largescale access of women (who would have traditionally taken on these tasks) to the world of work, as well as to the changing status of men and women in society. Several points can be made in relation to these changes:

– depending on how one sees a couple, whether one has a more modern or more traditional perspective, responsibility for waste is shared to a greater or lesser extent. The person who, within the couple, is in charge of shopping, cooking and even education will be a key player in the (anti-)waste behavior of the family;

– responsibility for shopping and meals does not lie solely with women, which has two consequences:

- a necessary distribution of tasks (who does what and when); the absence of discussion acting as a source of waste,

- a way of doing shopping that may differ among the two groups;

– the time spent on shopping is more limited (leading to a lack of menu planning, making one big shopping trip, potentially making online orders from the office without first checking what is still edible at home, etc.);

– men and women do not have an identical profile in food shopping. A study by Kantarmedia (2016)[13] shows, in particular, a certain duality in the shopping logic of men, more and more of whom are doing the shopping: they buy, for example, fewer *light* products but more processed products or ready meals and more organic and quality products; *in situ*, they also use the Internet to do their shopping more and tend to be "megaconsumers", meaning they buy in large amounts, thereby increasing the risk of waste. Women are more inclined than men to reduce food waste as they spend more time in the kitchen and are more exposed to the amounts wasted (Barr 2007).

Within the family and at home, the causes of waste can be numerous and cumulative. They can be linked to the buying process of the family during the consumption process.

4.3.1. *Managing shopping: a chronicle of foretold waste*

The first stage of all familial food consumption is buying food, whether it be raw (vegetables, fruit, meat) or pre-cooked. Lifestyle changes have led to significant evolution in the management of this stage: its modalities have changed and the factors to consider when going shopping have increased.

Finding solutions to reduce food waste requires identifying its sources and this demands answers to a litany of questions.

4.3.1.1. *Shopping: who does it? And how often?*

While practices have changed, it is still more often women (63%) *who are responsible for shopping*, especially in lower-income households (73% compared to 57% for higher socioprofessional categories)[14]. Having one person who is responsible for family food shopping potentially lowers the risk of waste by reducing duplicate purchases, while making this sole food provider responsible for planning meals and potentially other members of the family (Box 4.5). Shared responsibility requires a distribution of roles and shopping planning, which, if ineffective, increase the risk of

13 https://www.kantarmedia.com/fr/blog-et-ressources/blog/homme-pousse-chariot.

14 In-Store Media (2019). Available at: https://www.lsa-conso.fr/temps-frequence-motifs-d-achats-les-courses-passees-au-crible-etude,330781.

waste. Downloadable mobile apps exist, such as *note keep* for facilitating the distribution of shopping.

Bread constitutes 4% of family food waste (ADEME 2014). There are two underlying problems with this waste: 1) Who is buying it? 2) What can be done with leftover bread?

To the first question, one solution leaps out: decide who is responsible for buying the bread, stop moving the responsibility from one family member to another and/or between days of the week. This will avoid any days with either no bread or too much bread!

To the second question, the solution lies in the use of leftovers as the base ingredient for a recipe: French toast, bread pudding, breadcrumbs – there's no shortage of recipes!

Box 4.5. *Didn't anybody buy bread?*

Shopping frequency is also a factor in explaining the volume of household waste (Palatnik *et al.* 2014). Modern lifestyles have made the "big weekly shop" the option most often chosen by households[15]. Whether this is done online or in a supermarket, big shops confront the buyer with an enormous choice of that encourages impulse purchases, which, depending on the situation, replace or are added to the purchases initially anticipated. The consumer buys more and does not necessarily eat more. Moreover, this makes it more difficult to allocate a place in the week's menus for each food item purchased.

4.3.1.2. *How should we shop?*

Creating a shopping list (as eight out of ten French people do[16]) as part of meal planning (for certain days of the week), respecting it (16% stick to it strictly, largely for budgetary reasons[17]) and checking the contents of the fridge/cupboards beforehand naturally limit the risks of waste. The absence of meal planning, on the other hand, leaves room for an approximate idea of meals and improvisation (Box 4.6). However, the creation of a list suited to the needs of the family suggests specific advance knowledge of the family members that will eat each of the meals. This knowledge is highly dependent on the members of the family (age, activity), as well as the family's internal workings.

15 62% of French people state that they do one weekly shop and smaller occasional top-up shops.

16 In-Store Media (2019). Available at: https://www.lsa-conso.fr/temps-frequence-motifs-d-achats-les-courses-passees-au-crible-etude,330781.

17 *Ibid.*

Christine does the shopping at Carrefour. She dithers between two menus for Saturday evening dinner, lasagna or sushi, and finally chooses lasagna and buys the necessary ingredients. On Saturday afternoon, Christine prepares lasagna for five people (herself, her husband and their three children). Her 12-year-old daughter announces to her late that afternoon that she is invited to a sleepover at a friend's house. Meanwhile, her 17-year-old son decides to go the cinema with some friends. The 15-year-old has a sporting competition that afternoon and sends her a text to tell her that he will be back late and will pick up a sandwich to eat on the way. In the end, Christine and her husband decide to go to a restaurant; they will re-heat the lasagna tomorrow (it will be even better). Christine retrospectively congratulates herself for not having chosen sushi for the dinner that ended up being canceled… They only just escaped waste!

Box 4.6. *Lasagna or sushi? The difficulty of planning meals*

4.3.1.3. *Product choice: promotions and impulse buys*

A very large majority of consumers decide on at least some of their purchases, particularly food, *in situ*, that is, when they are in a store. There are multiple reasons for this: *making up for an oversight (66%), taking advantage of a promotion (61%) and buying a product that they want (54%)*[18]. Prices and promotions are also the purchasing triggers for many people. According to a study by In-Store Media, families with children demonstrate more specific characteristics in this context. They spend longer doing their shopping, they do it more often in very large supermarkets and are more likely to use a list, even if they then do not hesitate to discard it in favor of taking advantage of a promotion, trying a new product or because they feel inspired by another shopping cart.

In 2016, according to Crédoc[19], 72% of consumers state that they were attracted by a competitive price. While recourse to promotions is dependent on the purchasing power of households and the more or less optimistic economic climate of a country at a given time, this current practice has an immediate impact: it leads to families buying more to benefit from bargain prices. Most often, this unplanned surplus does not replace an item that the consumer eventually decides not to buy, but rather it is added to other purchases that were planned. As a result, either the family consumes more (there is more Coca-Cola® at home and so they drink more of it, which poses its

18 *Ibid.*

19 Hébel, P., Mathé, T., Vittori, A. (2016). Comment réagissent les consommateurs face aux promotions ? *Cahier de recherche*, C335, Crédoc, December 2016.

own problem of nutritional balance) or the product is put away, running the risk of being forgotten at the back of the cupboard until after its use-by date, thereby creating a new source of waste. Impulse purchases are similar. Drawn in, for example, by attractive packaging or a convincing demonstration of a food product in the store, the consumer buys products that are not on their list. These unplanned food products are added to those they already have at home, with a potential risk of waste (they eat the product bought on impulse today and the other products later but, in a domino effect, some products become unfit for consumption as they age in the interim).

4.3.2. *Sorting and storing groceries*

4.3.2.1. *How should shopping be stored?*

One of the sources of waste is linked to storage of family food purchases. The size of cupboards and the refrigerator affects how many of the products can be seen from the front. However, some of the sources of waste identified are due to a form of neglect, for example forgetting about foods that have become "too old". If it is not seen and not used, a product stored out of view can quickly be forgotten.

Reconsideration of storage spaces (different cupboards, refrigerator, freezer) after each shopping trip, in particular, (re)placing foods with the nearest expiration dates at the front, allows them to be used first (Box 4.7).

Different actions make it possible to prevent the growth of undesirable micro-organisms (harmful to hygiene and food quality) and limit food waste.

Device settings

1) Check the temperature of the refrigerator and freezer (+4°C and -18°C).

2) Check where the coldest part of the refrigerator is.

3) Respect the cold chain.

Food storage

4) Respect the temperatures mentioned on the labels.

5) Discard superfluous packaging (for example, cardboard on yoghurt packs) to limit food contamination.

6) Wrap raw and cooked foods separately (plastic wrap or foil).

7) Keep home-made meals in the refrigerator but eat them quickly.

8) Put the products with the nearest expiration dates at the front.

Refrigerator cleaning

9) Clean the refrigerator at least once per month.

10) De-ice the freezer and clean it at least once per year.

Box 4.7. *And if we took a little care of our refrigerator?*[20]

4.3.2.2. *'Use-by' and 'best before': what do they mean?*

Complete compliance with expiration dates on packaging also seems to be one of the causes of waste (Graham-Rowe *et al.* 2014) which in some cases could certainly be avoided.

Who has never reached as far back as possible on the shelf of a store to try to find the product with the longest expiration date? This common practice may seem justified if one is doing a large shop intended to last for a long time. Besides the fact that, in many cases, this is largely because of a desire to buy the "freshest" product (even when it is intended for almost immediate consumption), in reality this practice leads to the issue of food waste shifting from the family to the store, which is left with stock that is nearing expiry (Box 4.8).

Forgetting what one has already bought (which leads to duplicate purchases) or products at the back of the fridge is a significant source of waste.

Different brands of household appliances (for example, Samsung, Whirlpool, Haier, LG) are now proposing connected refrigerators that offer valuable help in reducing waste. Some, equipped with cameras, can be checked remotely via smartphone while the consumer is doing their shopping, in order to check which foods are already there. Some of them can send notifications to remind the consumer that a product is missing or that an expiration date is approaching, alert the consumer about the state of the products or to a temperature problem (allowing them to reduce energy costs) or provide recipe ideas on the basis of the products stocked! Others can issue an order for shopping with a partner brand. Lastly, the models have multimedia functions in order to reduce waste while making the experience enjoyable.

Box 4.8. *What about smart refrigerators?*

20 https://www.lci.fr/nutrition/dates-de-peremption-courses-comment-bien-ranger-son-frigo-pour-moins-gaspiller-2087261.html ; https://agriculture.gouv.fr/infographie-un-frigo-bien-range-des-aliments-mieux-conserves.

This practice doubtless has its origins in a relative lack of knowledge of the terminology used by producers. In that regard, the information provided by the French General Directorate for Competition Policy, Consumer Affairs and Fraud Control (DGCCRF)[21] clarifies matters:

– Use-by date:

> The use-by date indicates a hard limit. It applies to foods that are highly microbiologically perishable and that, because of this, are likely, after a short period, to present an immediate danger to human health. Set by professionals following aging tests, the use-by date is used for products such as charcuterie, fresh meat or cooked refrigerated meals. Foods with a use-by date are generally kept in the refrigerator. Their safety after the use-by date is only guaranteed if the storage conditions are respected.

– Best-before date:

> The use of a best-before date on the labels of food products is intended to inform the consumer of the date by which these foods retain their organoleptic, physical, nutritional, taste, etc. qualities. Provided that the packaging has not been tampered with, foods that have passed their best-before date can be consumed at no risk to the consumer.

A product that has passed its best-before date will have lost part of its smell, vitamins and organoleptic qualities, but it will not be unfit for consumption.

4.3.3. *During meals: What about waste at the table?*

According to ADEME, at least one product or part of a product, whether expired or damaged, is thrown away during the course of a week in all households. Most often, this concerns fruit or vegetables that are difficult to consume because of damage, but it can also refer to stale bread or products that have passed their expiration date (Masson and Gojard 2019). While, taken individually, families do not waste identically (volumes, types of products), the statistics provided by ADEME (2014) on the basis of the waste produced by 20 households makes it possible to identify the products wasted most (Figure 4.3).

---

21 DGCCRF: Direction générale de la Concurrence, de la Consommation et de la Répression des fraudes. Available at: https://www.economie.gouv.fr/dgccrf/Publications/Vie-pratique/Fiches-pratiques/Date-limite-de-consommation-DLC-et-DDM.

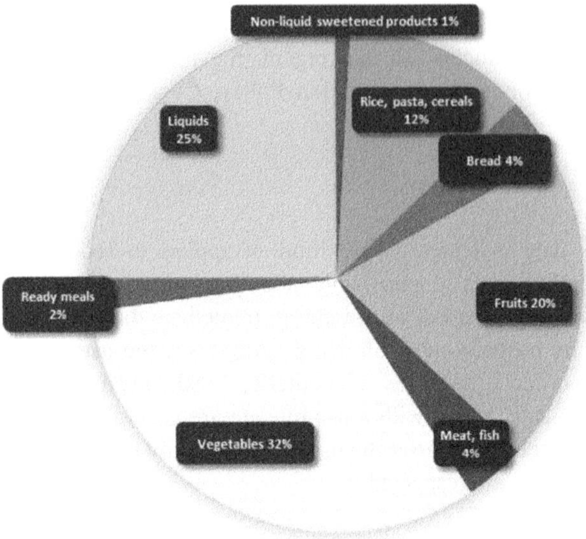

Figure 4.3. *The food products most wasted by households (source: study by ADEME (2014). For a color version of this figure, see www.iste.co.uk/dekhili/sustainable.zip*

Avoidable waste, corresponding to food thrown away when it is still consumable (64% of food waste).

Potentially avoidable waste, referring to food that consumers may or may not eat: cheese rind, radish tops, chicken skin, etc. (18%).

Unavoidable waste[22], which covers inedible food: bones, egg shells, etc. (18%).

Box 4.9. *The three categories of food waste according to WRAP*[23]

The leftover food on the plate at the end of a meal can, according to ADEME, be split into two groups (other categorizations are suggested such as those set out in Box 4.9):

– *so-called "selective" leftovers* that are related to the tastes, restrictions or habits of the consumer: fat from meat, cheese rinds, chicken skin or potato peel;

22 While food waste is defined as the act of discarding food items that are still edible during the consumption process (Le Borgne *et al.* 2018), unavoidable waste relates to the production of waste as a result of products not being consumable.

23 Source: Waste and Resources Action Program (WRAP), UK. Available at: https://www.wrap.org.uk/.

– *so-called "quantitative" leftovers* that are the result of an excess of uneaten food, which most often concerns vegetables and bread but can potentially also include other foods. This surplus can be explained by the person concerned becoming full more quickly than anticipated (pizza leftovers), inattention (the dregs in glasses or bowls) or impractical packaging (the remains of yoghurts or sauces). Taken together, these leftovers can constitute significant amounts of food.

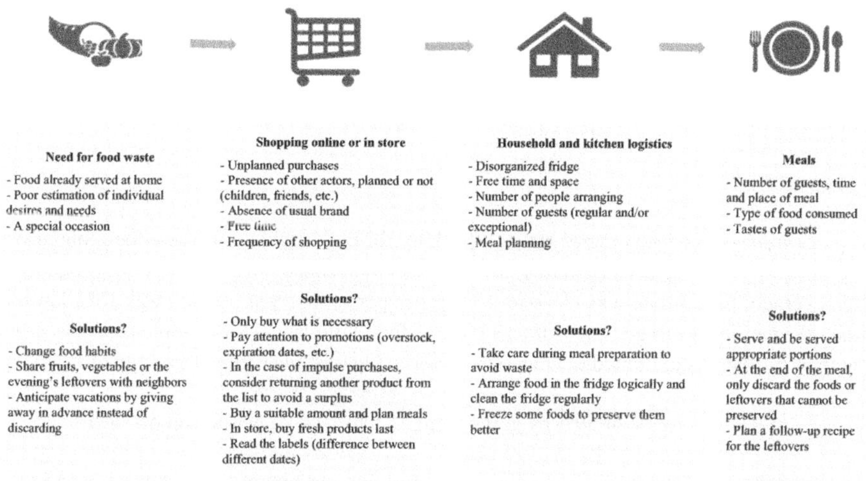

Figure 4.4. *Stages of consumption, sources of and solutions to the issue of food waste. For a color version of this figure, see www.iste.co.uk/dekhili/sustainable.zip*

4.3.4. *Proposed anti-waste solutions for each stage of consumption*

As we will see, the different stages of consumption can generate waste that can potentially be reduced through simple actions, as shown in the summary in Figure 4.4. The solutions proposed by ADEME in 2019 are also summarized in Box 4.10.

Between October 2018 and October 2019, 243 households (744 people) participated in an anti-waste operation for six months, led by three consumer associations[24] in partnership with ADEME.

The objective: provide them with daily support to change their behaviors and reduce their food waste.

24 The CSF (*Confédération syndicale des familles*) Bayonne, *Familles Rurales Normandie* (Rural Normandy Families), CLCV (*Consommation, logement, cadre de vie*) Hauts-de-France and Brittany.

The methodology: the respondents were all initially invited to evaluate the food waste in their household by weighing the discarded food stuffs (*liquids and solids*). Each household then chose three anti-waste actions to implement (*from nine that were proposed*) before proceeding to once again weigh the discarded food products. The items weighed each day concerned four categories of waste: solid and liquid food leftovers, as well as solid and liquid products still in their packaging.

The nine simple acts from which three had to be chosen were: 1) plan menus in advance; 2) make a shopping list and stick to it; 3) look at the expiration dates and appearance of products before buying; 4) ensure the proper storage of foodstuffs; 5) adapt to expiration dates; 6) prepare suitable quantities; 7) adopt an anti-waste kitchen; 8) serve suitable portions; and 9) keep the leftovers from meals.

The results

Awareness, the first crucial stage: especially for the weighing of waste food, but also because of its being divided into four categories (packaged/unpackaged x solid/liquid). On average, their waste was close to 25.5kg per person per year, which was 15% lower than the national average. The main source was food leftovers at 86%, while products still in their packaging amounted to only 14% of waste (against the national average of 23%).

A reduction in waste of 59% by going from 25.5 to 10.4 kg per person per year.

Liquids, the forgotten waste: a surprise for the participants, liquid foods are less visible and thus ignored as they are discarded in the sink, but account for a third of household waste (*milk, coffee, sauce, soda, soup, etc.*).

A sustainable and comprehensive change: three months after the end of the project, all of the households continue to carry out the actions, the most established of which were: "ensuring the proper storage of food" and "I look at the expiration dates and types of products to ensure they are suited to my needs".

The actions: the hardest – "I plan my menus in advance"; the easiest – "I keep the leftovers from my meals"; the most effective – "I prepare suitable portions".

Box 4.10. *Reducing waste: it is possible … and effective! A summary of a study carried out by ADEME in 243 households in 2019*[25]

25 https://presse.ademe.fr/2019/10/gaspillage-alimentaire-pres-de-250-foyers-ont-economise-lequivalent-de-21-400-repas-en-1-an.html.

4.4. Conclusion: What about the future?

The fight against waste is already being waged in France, particularly within families. Today it is a matter of going further by proposing new actions intended to develop positive attitudes among children, encourage their parents to adopt good behaviors and facilitate the implementation of solutions.

As regards families, in relation to the public authorities and associations, this concerns, for example, the establishment of educational projects about meal preparation, the proposal of cooking workshops to help to recycle leftovers, or the provision of practical information about learning to make compost or reducing the volume of waste when cooking.

When the family eats outside their home, such as in a restaurant, for example[26], it becomes a matter of intensifying awareness-raising activities among staff (of canteens, managers, chefs, etc.) and increasing quality guidelines, action guides, training, good practices[27] or even think tanks focusing on waste. This can only encourage researchers to carry out studies (such as on the use of the doggy bag, following the work of Achabou *et al.* (2018)) in order to find soft incentives (nudges[28]) to waste less, from the perspective of both families and caterers.

It is also, and this is a collective responsibility, a matter of involving children (the actors of tomorrow) in the fight against waste from a very young age by integrating them into educational projects that mix fun with sustainable development (work on the look of plates, for example; reducing noise in canteens to increase focus on meals, etc.). Furthermore, there should be encouragement for initiatives to launch cross-over projects that begin at school to be continued at home, in order to ensure that everyone is aware is that this concern affects us all.

In this regard, the role of the school, the major actor in the socialization of the child, is essential in ensuring that good behaviors are instilled (Nicolas-Hémar

26 15% of the main meals of French people (breakfast and dinner) are eaten outside the home, of which 52% are in commercial restaurants (with waste that rises to 211 g per person per meal). The food waste in mass catering is, on average, 167 g per person per meal. Source: Losses and waste in the direct delivery sector (catering and distribution), Ministry of Agriculture (2011).

27 The doggy bag practice became compulsory in French restaurants on July 1, 2021.

28 A nudge is a strategy that can take various forms. It aims to modify the cotextual elements that unconsciously influence the decision-making of individuals (Kahneman 2012) to lead them to make good decisions for their well-being and/or that of society. An example of a nudge would be reducing the size of plates themselves; this naturally leads to smaller quantities of food being served.

et al. 2013). The law on combating food waste (2016) now requires schools to inform and educate their pupils on the fight against food waste. Indeed, because there are many children and teenagers that have breakfast there each day, school canteens are significant sources of food waste (on average, according to ADEME (2019)[29], 110–130 g per person per meal in elementary schools, 179–200 g per person per meal in junior high and high schools). While the losses are not solely due to waste from children (other factors include the overestimation of the amounts to prepare, changing numbers, overly large portion sizes, unattractive quality and presentation of foods, limited time to eat, etc.), the role of young eaters is real and raising their awareness is absolutely necessary.

The Ministry of National Education strongly recommends mentioning responsible practices at the youngest possible age by means of cross-over topics that must be included in teaching practices[30]. The aim is to raise awareness of sustainable development in general and, in particular, to guide pupils towards "good practices". Waste linked to food is therefore one of the seven pillars of these recommendations. As it happens, there is increasing integration of the issue of responsibility in daily school activities (choirs on the issue of pollution, anti-waste processes in the canteens, lessons about the planet, voluntary reports on changing climatic conditions), albeit with difficulties and paradoxes (increased awareness of waste, but reduced attention to sorting waste owing to a lack of time or staff training, for example) (Clauzel *et al.* 2016). There are numerous initiatives emerging from different associations, following the example of the *Graine Île-de-France* educational project (and the associated white paper[31] containing anti-waste educational tools: anti-waste tasting, anti-waste recipes by a chef, an anti-waste kitchen made of recycled cardboard, shows with local entertainers, easy recipes for quick meals).

Focusing solely on food is not enough to stop global waste because food waste is only one aspect of the ongoing environmental issues and because becoming experts in reducing food waste does not necessarily turn a family into responsible consumers. It is therefore also important to raise families' awareness of other sources of waste in order to make greater headway in the fight against all forms of waste and for the progress made to be visible and real on a global scale: this includes energy consumption, recycling toys and clothes, using clean transport, recovering computer and digital equipment and more.

---

29 https://optigede.ademe.fr/fiche/lutte-contre-le-gaspillage-alimentaire-en-etablissements-scolaires.

30 https://eduscol.education.fr/cid91752/nourrir-planete-nourrir-humanite.html.

31 http://www.graine-idf.org/partenariats/annuaire-gaspillage-alimentaire.

4.5. References

Achabou, M.A., Dekhili, S., Tagbata, D. (2018). Mieux comprendre les difficultés de développement du Doggy Bag en France : une analyse par l'approche des représentations sociales. *Décisions Marketing*, 92, 53–76.

Barr, S. (2007). Factors influencing environmental attitudes and behaviors: A UK case study of household waste management. *Environment and Behavior*, 39(4), 435–473.

Charry, K.M. and Demoulin, N.T. (2012). Behavioural evidence for the effectiveness of threat appeals in the promotion of healthy food to children. *International Journal of Advertising*, 31(4), 773–794.

Clauzel, A., Guichard, N., Riché, C. (2016). Papa, pourquoi tu jettes cette tomate ? Elle est toujours bonne, c'est la maîtresse qui l'a dit. Dynamique décisionnelle familiale et apprentissage réciproque des comportements antigaspillage. *32ème Congrès international de l'Association française du marketing*, Lyon, France, 18–20 May.

Clauzel, A., Guichard, N., Riché, C. (2017). C'trop la hess de gaspiller, déter pour arrêter ? Ce que les adolescents pensent du gaspillage alimentaire. *33ème Congrès international de l'Association française du marketing*, Tours, France, 17–19 May.

Clauzel, A., Guichard, N., Riché, C. (2020). Perceptions du gaspillage alimentaire par les adolescents. Les apports de la théorie de l'attribution causale. *2ème Journée sur le Marketing et Développement durable (JMDD)*, Strasbourg, France, 6 February.

Desbordes, A. and Botavant, M. (2005). *La soupe à tout !* Éditions Nathan, Paris.

Ekstrom, K.M. (1995). Children's influence in family decision making: A study of yielding, consumer learning, and consumer socialization. BAS Gothenberg University, Gothenberg.

Freeman, D. and Brucks, M. (2002). Drugs, alcohol, and tobacco use prevention efforts. Benefits and challenges of targeting young children. In *Children: Consumption, Advertising and Media*, Hansen, F., Martenses, A., Tufte, B.D. (eds.). Business School Press, Copenhagen.

Gollety, M., Guichard, N., Ayadi, K. (2012). La consommation dans la famille : des influences de nature différente. In *Kids Marketing*, Brée, J. (ed.). EMS, Caen.

Graham-Howe, E., Jessop, D.C., Sparks, P. (2014). Identifying motivations and barriers to minimising household food waste. *Resources, Conservation and Recycling*, 84, 15–23.

Grønhøj, A. and Thøgersen, J. (2012). Action speaks louder than words: The effect of personal attitudes and family norms on adolescents' pro-environmental behavior. *Journal of Economic Psychology*, 33(1), 292–302.

Kahneman, D. (2012). *Thinking Fast and Slow*. Penguin Books, New York.

Kelley, H.H. (1973). The processes of causal. *American Psychologist*, 28(2), 107.

Kral, T.V. and Hetherington, M.M. (2015). Variability in children's eating response to portion size. A biobehavioral perspective. *Appetite*, 88, 5–10.

Le Borgne, G., Sirieix, L., Costa, S. (2018). Perceived risk of food waste: Influence on consumer attitudes towards and choice of sales promotions. *Journal of Retailing and Consumer Services*, 42, 11–21.

Masson, M. and Gojard, S. (2019). Le gaspillage alimentaire dans les foyers français. *Cahiers de nutrition et de diététique*, 54(4), 240–246.

Nabec, L. and Roux, D. (2016). *Protection des consommateurs : les nouveaux enjeux du consumérisme*. EMS, Caen.

Nicolas-Hemar, V., Ezan, P., Gollety, M., Guichard, N., Leroy, J. (2013). How do children learn eating practices? Beyond the nutritional information, the importance of social eating. *Young Consumers*, 14(1), 5–18.

Palatnik, R. (2014). Greening household behaviour and waste. OECD Environment Working Papers, 76.

Pechmann, C., Levine, L., Loughlin, S., Leslie, F. (2005). Impulsive and self-conscious: Adolescents' vulnerability to advertising and promotion. *Journal of Public Policy & Marketing*, 24(2), 202–221.

Schill, M. and Fosse-Gomez, M.H. (2014). La place de l'enfant dans le comportement écologique de la famille : une approche par le mime. *13ème Journées Normandes de Recherche sur Consommation*, Rouen, France, 27–28 November.

Stefan, V., Van Herpen, E., Tudoran, A.A., Lähteenmäki, L. (2013). Avoiding food waste by Romanian consumers: The importance of planning and shopping routines. *Food Quality and Preference*, 28(1), 375–381.

5

The Packaging-free Product Market: A Renewal of Practices

Maud DANIEL-CHEVER[1], Élisa MONNOT[2], Fanny RENIOU[1] and Lucie SIRIEIX[3]

[1] *CREM-CNRS, IGR-IAE Rennes, University of Rennes 1, France*

[2] *THEMA, CNRS, CY Cergy Paris University, France*

[3] *UMR MOISA – CIRAD, University of Montpellier, CIHEAM-IAMM, INRAE, Institut Agro-IRD, Montpellier, France*

5.1. Introduction

While it remains difficult to establish sustainable practices in the daily lives of individuals and their lifestyles, some trends in consumption seem to be gaining ground. This can be seen in the success of fair trade products[1], the growth of the market for organic products[2] and the increase in sales outlets for local products[3], all driven by arguments linked, or not, to sustainability (Daniel and Sirieix 2012). These modes of sustainable consumption also include the purchase of unpackaged products (Monnot and Reniou 2012), a new way of buying that seems to have had a positive response among consumers. In France, the market on which this chapter focuses, the sales outlets for packaging-free products are multiplying, the range of products sold without packaging is growing, producers are demonstrating a great capacity for innovation and supply chains are being organized. Célia Renesson,

1 See: https://www.commercequitable.org/le-commerce-equitable/quelques-chiffres.

2 See: https://www.agencebio.org/wp-content/uploads/2019/06/DP-AGENCE_BIO-4JUIN2019.pdf.

3 See: https://www.lsa-conso.fr/le-local-prend-son-essor,316081.

Marketing for Sustainable Development,
coordinated by Sihem DEKHILI. © ISTE Ltd 2021.

president of Réseau Vrac (a network focused on packaging-free products), estimates that in France in 2018, the market for packaging-free products had a turnover of 850 million euros, compared to only 100 million euros in 2013[4]. While the distribution of packaging-free products is not new, as it was the predominant distribution method in the 1950s, this market is nonetheless undergoing a renewal with strategic and operational upheavals. Packaging-free products are no longer driven by a group of consumers with no other alternatives, but by consumers who, faced with an abundance of (over-)packaged products, are making the choice to buy products without packaging, without plastic and so on.

Prior to the 1960s, the most typical method was to distribute unpackaged products. Between 1945 and the oil crisis of 1973, societies underwent numerous changes: less time was spent on domestic tasks, such as food shopping and meal preparation, and household equipment developed considerably, with the advent of the refrigerator and the freezer, which changed methods of shopping, storing and consuming. At the same time, agrifood industries were offering a growing variety of packaged and ready-to-use products that could be bought more and more easily in large and medium-sized stores. While the products could otherwise, for the most part, be felt, seen and touched before being bought (Ferrières 2002), today they are often "hidden" in packaging. In developing this packaging of products, merchants have led consumers to change the way they choose products. Packaging has become a commercial interface (Barré and Heilbrunn 2012) that, like the brand and its advertising, acts as an intermediary between the consumer and the product (Cochoy 2002). The products are acquired for their content, but also because of the container and its brand.

Recently, with the return of packaging-free products, packaging has no longer constituted either a barrier to the product or a factor in the decision to buy it. Consumers must therefore get used to new ways of consuming (Beitzen-Heineke *et al.* 2017). Some countries have remained rooted in the mode of packaging-free product consumption. In Asia, for example, where mass distribution has developed, the range of unpacked products sold is large: rice, for example, needs to be touched and smelled (Figuié and Bricas 2016). In Canada and Italy, enthusiasm for packaging-free products consumption has been translated into initiatives, such as the online grocery Boco Boco and the brand Negozio Leggero. In France, this market is still a niche one, and the products purchased by households are, for the most part, still industrialized and packaged.

What exactly does the term "packaging-free product" include? In France, there is no official definition. According to Réseau Vrac, the term "vrac", meaning "packaging-free products", indicates "the retail sale of products that are not

4 See: https://www.lopinion.fr/edition/economie/celia-rennesson-reseau-vrac-commerce-en-vrac-partie-reponse-reduire-199525.

pre-packaged, using a self-service system, and in a desired quantity in a reusable container"[5]. Although some products are not yet accessible, owing to technological or regulatory restrictions, the range of unpackaged products available is growing all the time, in both the food sector (oils, yoghurts, etc.) and the non-food sector, particularly cosmetic products (solid shampoo). This growing offer is the combined result of increasing consumer demand and suppliers' greater capacity for innovation.

With the help of the analysis of a study carried out through interviews with consumers and experts (D. Onraita co-founder of the Day by Day network; M. Cadic of the Mamie Mesure network and C. de Boisse of the Briochin brand)[6], this chapter aims to respond to two major questions about the (re)birth of the packaging-free product market.

– *How can we characterize the consumption of packaging-free products*? Where does the enthusiasm for packaging-free consumption come from? Who is buying packaging-free products? What are the drivers and limitations of packaging-free consumption?

– *What are the proposals for the packaging-free product market*? What modes of distribution have been adopted? What are the available marketing provisions? What does packaging-free consumption change for suppliers/retailers?

This also leads us to respond, in conclusion, to the question of the capacity of consumption of packaging-free products to participate in the revival of sustainable development, by broadly investigating the links between packaging-free products and waste, the "Zero Waste" approach and sustainability.

5.2. The characteristics of packaging-free consumption

5.2.1. *Where does the enthusiasm for packaging-free products come from?*

The growth of the power of packaging-free consumption has occured in tandem with increasing media coverage of the environmental and health impact of packaging, particularly plastics.

This media coverage mainly concerns an *environmental assessment*. Each year, 300 million tons of plastics are produced worldwide, of which half, primarily plastic

5 See: https://reseauvrac.org/.

6 All of the data collected using these different methods is used to serve the arguments outlined in this chapter. For example, the experts were questioned about their vision of packaging-free consumption and its development in France, their view of consumers, the restrictions imposed by this method of consumption/distribution, etc.

packaging, end up adrift in the ocean[7]. French households produce 568 kg/year of waste[8], which constitutes significant quantities. The solution currently being promoted is the sorting and recycling of packaging. However, this remains limited as not all packaging is recyclable and, for those that are, the recyling process is not infinite and remains costly from an energy perspective. The ideal solution therefore remains as the elimination of packaging at the source.

Media coverage then involves a *health assessment* by challenging the harmlessness of plastic packaging (bisphenol A, plasticizers, styrenes and other endocrine disruptors). According to the Center for International Environmental Law (CIEL), plastic represents a risk to health on a global scale, as it has a harmful effect when it is manufactured, used, processed as waste and dispersed into the environment[9].

This environmental and health context has helped to raise awareness of the risks associated with packaging and the birth of a now widely shared movement, the *"Zero Waste" approach*. This movement, driven by initiatives from different communities and individual and insitutional actors, has a strong impact on society and is directly participating in increasing consumer enthusiasm for packaging-free consumption. With regard to *community initiatives*, the association Zero Waste France, created in 1997, promotes a zero waste approach that prioritizes the reduction of packaging at the source. This conists of combating "disposable" items and promoting alternatives to single-use products used on a daily basis, which are responsible for generating significant quantities of waste. When it comes to *individual initiatives*, a number of citizens have become figureheads of this movement. They include Béa Johnson, who has worked to improve the dynamism and dissemination of the movement. Her 2013 book on Zero Waste, which provided household tips for decluttering the home, is a bestseller and has been translated into 27 languages[10]. Moreover, since 2018, people worldwide have been organizing "plastic attacks", which consist of going to a supermarket to do their shopping and leaving all of their superfluous packaging there. This movement created a domino effect, as it then led to numerous actions in different countries: *the aim is to raise customer awareness of the overconsumption of plastic, as well as to inform*

7 See: https://www.liberation.fr/planete/2019/11/14/chaque-annee-8-millions-de-tonnes-de-plastique-finissent-dans-les-oceans_1763117.

8 Que contient notre poubelle ? ADEME infographic, July 2019.

9 Plastic and health: The hidden costs of a plastic planet. CIEL report, February 2019 [Online]. Available at: https://rethinkplasticalliance.eu/wp-content/uploads/2019/02/Plastic-and-Health-The-Hidden-Costs-of-a-Plastic-Planet-February-2019.pdf.

10 See: https://www.lemonde.fr/festival/article/2017/09/05/bea-johnson-le-zero-dechets-radical-chic_5181353_4415198.html.

supermarkets of the uselessness of all of this packaging[11]. With regard to *institutional initiatives*, in 2014, the city of Roubaix launched a Zero Waste initiative without precedent in France: 500 families had their consumption monitored and were encouraged to reduce their waste[12]. At the national level, Brune Poirson, French Secretary of State for the Environment, highlighted 50 measures in 2018 in the *Roadmap for the Circular Economy*[13], which included restrictions on single-use packaging. Lastly, the European Union approved a ban on single-use plastic from 2021.

The media coverage of the Zero Waste movement and the institutional measures taken thus support the promotion of the market for packaging-free products. Where it was formerly secret and lacking data, nowadays this market is growing and becoming ever more accessible and visible for consumers, whatever their usual sales outlet may be (large and medium-sized stores, specialized shops, online sales outlets or markets). Packaging-free consumption constitutes a real opportunity for suppliers and retailers, provided that they take consumers and their character, motivations and fears into account, which we will go on to detail.

5.2.2. Consuming packaging-free products

The current state of packaging-free consumption in France leads us to a first assessment: the *heterogeneity of the data* from studies. The given percentage of French people who have bought packaging-free products in the last 12 months ranges from 37% (Nielsen 2018[14]) to 47% (Kantar 2018[15]). All of these studies do, however, tend to lead to the same conclusion: the consumers of packaging-free products are gradually becoming less of a minority. However, given the frequency with which these products are bought, the trend needs to be put into perspective: the average number of purchases per year is still low (11.3%), and only 24% of buyers of packaging free products buy these products at least once per month[12].

11 See: https://www.lefigaro.fr/conso/2018/06/02/20010-20180602ARTFIG00075-la-france-vit-sa-premiere-plastic-attack-contre-le-suremballage.php.

12 See: https://www.roubaixzerodechet.fr.

13 See: https://www.ecologique-solidaire.gouv.fr/feuille-route-economie-circulaire-frec.

14 See: https://www.nielsen.com/fr/fr/insights/news/2019/bulk-buying-habit-that-settles-in-the-french.html.

15 L'Observatoire du rayon vrac 2019 [Online]. Available at: https://www.olivierdauvers.fr/2019/10/18/en-libre-telechargement-lobservatoire-du-rayon-vrac-2019/.

Is it possible to create a *composite image of consumers* of packaging-free products? Studies show that there are few sociodemographic differences within this group, apart from those aged under 35, who appear to be keener on packaging-free shopping than the average French person (47.6% versus 46.5% among those over 65[12]). Consumers of packaging-free products are differentiated more by their motivations for buying them than by their profession or level of education. In general, they are more environmentally aware and show this in their shopping habits: they buy organic, fair trade and local products in higher-than-average quantities and they read packaging carefully (see footnote 27). D. Onraita (Day by Day) has identified three customer profiles for those buying packaging-free products in specialized shops, in accordance with how often they buy and their motivations:

– the biggest group are the "curious", who have no initial desire to change their consumption habits, but say to themselves "well, why not" and thus need to be reassured;

– then there are the "switchers", who have a more distant attitude to packaging-free consumption and are looking for products equivalent to those they normally buy;

– and lastly there are the "converts", who are looking for a complete range of products that allow them to limit packaging as much as possible.

Packaging-free consumption would therefore make it possible to respond to the needs of a varied audience: customers anxious to consume products that are healthier and more environmentally-friendly; families that are looking to educate their children through these modes of consumption; older people who appreciate the opportunity to connect and are nostalgic for the shopping habits of days gone by; and young people with budgetary restrictions[16]. Understanding this broad range of customers necessarily includes studying the motivations and obstacles associated with packaging-free consumption.

5.2.3. *Motivations for and obstacles to packaging-free consumption*

Regarding the *motivations for choosing packaging free consumption*[17], our study allows us to identify five main motivations, as indicated in Figure 5.1. These motivations are related to protecting the environment (reducing packaging), being

16 Faut-il s'emballer pour le vrac ? *LSA*, 2419, June 30, 2016.

17 To identify consumer motivations, a sample of 124 respondants were asked the following question: "What are your motivations for choosing packaging-free consumption?" A total of 146 reasons were given, which can be grouped around five topics.

able to choose the desired quantity (waste limitation), stocking up on certain types of products (buying cereals, grains, dry products, etc.), saving money (accessible prices) and looking for practicality in consumption (ease of use, simplicity, speed).

The results of our study combine with the motivations identified in the Kantar study (2018), which were as follows: "to have the precise amount I need" (54%); "to buy a greater variety of things in smaller quantities" (39%); "because these are healthier and more natural products" (17%); and "because it's cheaper" (15%), as well as "to use less packaging" (53%). Beyond concerns related to reducing unnecessary packaging and food waste, the economic benefit can be a reason to opt for packaging-free consumption. Packaging-free consumption can be a means of responding to the consumer need for variety, as it can facilitate certain types of impulse purchase by allowing consumers to taste new products, in smaller quantities and at a lower cost[18].

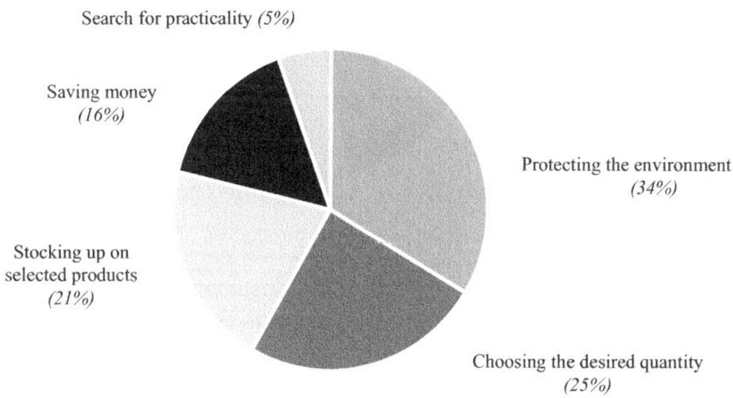

Figure 5.1. *Motivations for packaging-free consumption (n=146). For a color version of this figure, see www.iste.co.uk/dekhili/sustainable.zip*

With regard to *obstacles to packaging-free consumption*[19], our study shows that the most mentioned obstacles are the logistical organization it requires (consumers have to bring their own containers), the prices (which were perceived to be too high), the general accessibility of packaging-free consumption (either sales outlets

18 Faut-il s'emballer pour le vrac ? *LSA*, 2419, June 30 2016.

19 To identify consumer impediments, a sample of 124 respondants were asked the following question: "What are the impediments to you choosing packaging-free consumption?" A total of 168 reasons were given, which can be grouped around 11 topics.

that do not offer packaging-free consumption or these sales outlets were far away), the lack of a habit and ignorance of this type of consumption, the perceived poor quality of products, hygiene and a lack of choice. The lack of information on products, difficulties in measuring quantities and perceived disorderliness around packaging-free arrangements, as well as the views of others in stores, can also explain, to a lesser extent, consumer reticence to packaging-free consumption (Figure 5.2).

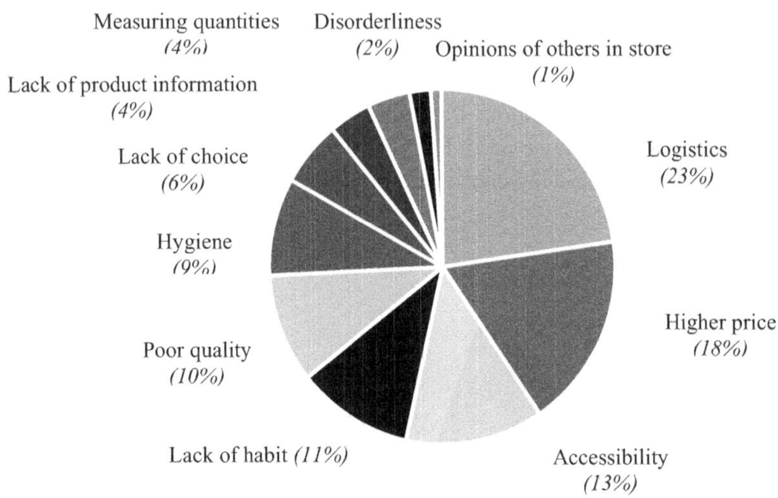

Figure 5.2. *Obstacles to packaging-free consumption (n=168). For a color version of this figure, see www.iste.co.uk/dekhili/sustainable.zip*

These obstacles are, to certain extent, in accordance with those mentioned in a study by the Observatory on the Circular Economy in France (ObSoco) (2019)[20]: this study noted the reduced offering and the lack of brand engagement as the main reasons for not adopting packaging-free consumption (60%). The lack of practicality (49%), prices perceived as being higher (39%), hygiene (35%) and a lack of product information and safety (33%) were also mentioned. These obstacles have also been confirmed by experts:

> The obstacles always relate to two things: time and money. For a transition to packaging-free consumption, there is an initial obstacle linked indirectly to money: taking a risk on the product. [Customers say] "I don't know the products which are sold because the big brands

20 Observatoire du rapport des Français à l'économie circulaire. Study, *L'ObSoCo*, September 2019, p 15.

> aren't involved" [or talk about concerns related] to hygiene… Brands therefore have to get involved to build confidence and points of reference, and regulations prevent suppliers from ensuring faultless traceability and hygiene. The second restriction is time. Packaging-free consumption currently seems extremely time-consuming, which requires a new kind of organization. (D. Onraita, Day by Day)

It is thus possible to identify *five major obstacles* to packaging-free consumption from these studies.

Accessibility of provision: even though the number of stores offering packaging-free consumption is continually increasing[21], the stores most often visited by consumers do not offer these products. Individuals also have their own brand preferences and these cannot be sold unpackaged, which limits the reach of packaging-free consumption.

Practicality and logistics: consumers must plan all logistical organization themselves (remembering to bring their own containers, transporting them, storing them). Packaging-free consumption can therefore be associated with a lack of ease, as managing these containers can be perceived to be complicated and time-consuming.

Product prices: packaging-free consumption can have a relatively ambiguous image of "price" among consumers. On the one hand, it can be associated with low-cost consumption. This low-cost image has been developed over the years in retail chains such as Auchan with a "self-discount" concept; this is targeted at customers wishing to buy small portions of staple products in order to minimize their expenditure[22]. On the other hand, it can also be associated with "less controlled" and thus potentially more expensive consumption, particularly because of products being priced by weight and not by unit.

Hygiene and perceived quality: while there are regulations in this area, the hygiene of handling products, the equipment used and locations concerns consumers. The laws of contagion are thus applied to packaging-free consumption (Rozin and Nemeroff 2002): some individuals may have the impression that some

21 In late 2018, France Inter published an interactive map of 280 shops with 100% packaging-free products, by collaborating with Internet users who could propose brands to identify. See: https://www.franceinter.fr/societe/exclusif-la-carte-des-epicerie-100-vrac.

22 See: https://www.20minutes.fr/economie/1623903-20150605-auchan-poursuit-developpement-rayons-vrac.

packaging-free products have been touched by people carrying bacteria and are therefore contaminated (Argo *et al.* 2006).

Lack of product information: consumers may sometimes feel lost in the face of packaging-free products as they do not offer as much information as their packaged equivalents. The latter have a brand that offers a guarantee of quality and a reassuring safety; with packaging-free products, the brand disappears and the product information is reduced or, at least, conveyed by means other than packaging.

Having presented the characteristics of packaging-free consumption (the reasons for its emergence, the types of consumers concerned and the motivations/obstacles associated with this mode of consumption), we will now turn to the offerings available on the packaging-free product market.

5.3. Offerings on the packaging-free product market

5.3.1. *Positioning strategies of packaging-free product suppliers*

Packaging-free consumption is currently primarily the preserve of organic stores (where 51.5% of consumers say they buy packaging-free products), but large and medium-sized stores, which are increasingly interested in this market, account for 40% of sales[23]. The number of small grocery stores selling packaging-free products is also growing rapidly in France: there were 15 of them in 2018, and 300 by the end of 2019[24]. Other channels, which are much more than anecdotal, are also appearing: mobile sales outlets at markets, drive-ins and even home deliveries through online sales.

By reviewing the existing initiatives and different criteria, varied positioning strategies emerge[25]. These are as follows.

The importance of the range of packaging-free products in comparison to the range of packaged products: from stores specializing in 100% packaging-free products, such as Day by Day, in hypermarkets, whose packaging-free range is only

23 Observatoire du rayon vrac 2019 [Online]. Available at: https://www.olivierdauvers.fr/2019/10/18/en-libre-telechargement-lobservatoire-du-rayon-vrac-2019/.

24 See: https://www.lopinion.fr/edition/economie/celia-rennesson-reseau-vrac-commerce-en-vrac-partie-reponse-reduire-199525.

25 White paper. *Gérer un espace de distribution en vrac en 2017* [Online]. Available at: http://vracoop.fr/guide-du-vrac/.

a part of their offering, to shops selling organic products, there is a wide range of choices. These choices have significant differences in terms of logistics, merchandising and sales assistance.

The range of food and non-food products: the choice and the range of products lead stores to develop very different strategies, often guided by the needs of their regular customers, as well as by logistical and financial constraints. "Having a food shop is a tremendous amount of work. We invest a lot to make a profit on food and it remains low. With non-food products, the margin is much more interesting and is a lot more practical to manage as it involves small amounts, small things..." (M. Cadic, Mamie Mesure).

Price positioning: the choice of which products are sold without packaging is a question of brand policy. Even if the removal of packaging makes it possible to reduce the costs of making a product, the sale volumes are not always high enough to guarantee their competitiveness. Therefore, retailers sometimes reduce their margins, as, for example, Biocoop does for certain products[26]. This is a strategic choice as the prices remain a determining criterion in food purchases.

Level of environmental engagement: this can take the form of three kinds of actions. Firstly, it involves determining whether the products sold unpackaged are organic or not. Some brands, such as Biocoop in France, certify all of their packaging-free products as "AB" ("*Agriculture biologique*" – organic agriculture) in accordance with their positioning. In large and medium-sized stores, 57% of products sold unpackaged are included in the range of organic products, often linking packaging-free consumption to organic issues more broadly[27]. According to our study, packaging-free products are also often associated with organic or environmentally-friendly products (see text Box 5.1). The retailer Carrefour is one of the retailers in which packaging-free products are particularly established and 96% of its packaging-free products are included in the range of organic products. Some retailers specializing in packaging-free products are also choosing to focus on organic products, even if that means not being to cater to all needs, such as, for example, Biovrac in Biarritz or the Juste Bio website. Secondly, there is the question of whether retailers will accept customers coming with their own containers, or whether they will provide them. While, in large and medium-sized stores, consumers are rarely encouraged to come with their own containers, the majority of stores specializing in packaging-free products (such as Day by Day) suggest it. The

26 See: https://www.ademe.fr/sites/default/files/assets/documents/86337_etude-vrac-12-2012.pdf.

27 Observatoire du rayon vrac 2019 [Online]. Available at: https://www.olivierdauvers.fr/2019/10/18/en-libre-telechargement-lobservatoire-du-rayon-vrac-2019/.

e-commerce platform Loop uses another function, online ordering, and delivers the products ordered in a bag with different compartments and in receptacles designed to be reused. Thirdly, chains must establish whether they are happy to simply sell packaging-free products, or whether they will offer broader services to support consumers in their Zero Waste efforts. The shop Franprix Darwin, opened in Paris in 2017, goes beyond simply selling packaging-free products and offers its customers solutions to limit their waste packaging, as well as a kitchen space equipped with dishes to have a Zero Waste lunch break.

> In order to understand the representations associated with packaging-free consumption, our study also included the following question, which the same 124 respondants were asked: "When I say packaging-free consumption, what are the three words that come to mind?" Eight major representations were identified, corresponding to 221 mentions of 353: "environmentally friendly", "choice", "dry products", "organic", "economical", "mess", "groceries" and "expensive".

Box 5.1. *The major images of packaging-free consumption*

Figure 5.3. *Word cloud relating to packaging-free products. For a color version of this figure, see www.iste.co.uk/dekhili/sustainable.zip*

Proximity: the proximity highlighted by some retailers takes two forms. Firstly, this can mean proximity to customers. According to S. Leflond, "80% of customers are within 500 meters of a store. [The predominant criterion identified by customers is] proximity. This indicates that packaging-free product companies must favor the best possible location and organize their distribution processes around their customers"[28]. This is done by, for example, Day by Day, whose value proposition centers on their proximity to customers, allowing them to do their shopping from day to day ("Day by

28 White paper. *Gérer un espace de distribution en vrac en 2017* [Online]. Available at: http://vracoop.fr/guide-du-vrac/.

Day, your grocery store for packaging-free products, in your neighborhood"). Secondly, this concerns proximity to suppliers in order to acquire local goods. For example, Les Petites Graines, a mobile grocery in Finistère, prioritizes supplies from local producers.

Regardless of the positioning strategy taken, the retailers of packaging-free products structure themselves around two main groups of actors: on the one hand, organic retailers and retailers specializing in packaging-free goods, experts in the sector who belong to the Réseau Vrac network (text Box 5.2); and, on the other hand, large and medium-sized stores, which are gaining market share, but are struggling to gain legitimacy owing to a lack of knowledge and poor management of their range (e.g. limited selection, unappealing look, few reusable containers). One solution could be for these two types of actors to pool their knowledge: the sales strength of one and the expertise of the other. This is, in fact, the idea implemented in the store Cora in Ermont, in which the Day by Day brand has established itself, with four employees in 100 m^2 of sales space, by faithfully reproducing its concept. Previously, Cora had only sold around forty packaging-free products in silos in the health food aisle; this new set-up allows it to benefit from a broader range (of food and non-food products) and the packaging-free produce experts benefit, for their part, from the appeal of the hypermarket[29].

Created on March 10, 2016, Réseau Vrac is an association of professionals based in Paris, dedicated to promoting the sale of packaging-free goods in France and around the world. Its aim is to participate in structuring and developing this new market. The values and commitments of Réseau Vrac are to reduce disposable packaging and food waste, create social connections and combat economic inequalities, develop jobs and the local economy, and support sustainable and responsible consumption[30].

Here follows an extract from an interview with Maud Cadic, a trainer in the Réseau Vrac association, and co-founder of the grocery store Mamie Mesure.

How is Réseau Vrac organized?

"It's an organization that emerged from Zero Waste France because there was a lot of demand from people who wanted to go to packaging-free stores and who contacted Zero Waste [...]. Four years ago, there were 50 members of the network and today there are 1,500. We're also developing a presence abroad as we're about to launch Réseau Vrac Belgium. It's happening very fast! So this requires us to make good decisions and to make them quickly.

29 See: https://www.olivierdauvers.fr/2019/08/09/le-premier-corner-de-lenseigne-vrac-day-by-day-chez-cora/.

30 See: https://reseauvrac.org/.

Réseau Vrac is made up of an administrative council, a director, a training team, service providers and members. The members are both small and large stores, and project leaders and suppliers. The headquarters are in small premises in Paris. Until recently, the association operated on a purely voluntary basis. However, the network is becoming more structured and the increase in member numbers is making it possible to pay some employees."

What are the aims of Réseau Vrac?

"The democratization of packaging-free consumption and support for professionals to avoid everyone developing on their own and doing different things, so that we're all moving in the same direction and in accordance with what the authorities [...] are telling us to do. There's no obligation to join the network though! People can launch their own shops by themselves. It's just that the network provides support, particularly legal support. For example, if a member receives a fine because of an error or a labeling issue, they can bring their problem to Réseau Vrac at the national level, to inform the authorities and get the right information out to everyone. There is also monitoring carried out by the legal directorate, a well-stocked and continually updated media library with information about the sector and, for project leaders, there is access to specialized training."

The rapid growth of the network shows the significant need for support for professionals in a sector where production, logistical and distribution practices are constantly changing and improving, and in a sometimes uncertain legal context. The challenges will remain the same in the coming years: confronting this growth and carrying out a discussion with a single voice, while taking into account the diversity of members, and considering whether or not to integrate the major actors in distribution.

Box 5.2. *Réseau Vrac, a network to support professionals in the sector*

5.3.2. *The offering proposed and the range of products*

5.3.2.1. *Specificities of packaging-free consumption and the legal framework facilitating its development*

Before presenting the range of products sold unpackaged, it should be noted that these products have specific features compared to equivalent packaged products. First of all, packaging-free selling involves the removal of brands. Brands have become a means of helping consumers to choose which pacakaged products to buy, with the brand providing a guarantee, especially in terms of the quality of the product. There is also the question of the specific legal regulations for packaging-free products. In principle, all food products can be sold without packaging, but there are some exceptions, such as frozen products. French law and, in particular, the DGCCRF[31] specify that the information that must be provided on food products themselves or in the immediate vicinity on a sign is: the sales name of the product (description), its physical state (frozen, for example), its sales price (by unit or weight), its composition (list of ingredients when it is a processed product) and, where relevant, the presence of allergens (nuts, gluten), as well as any potential ionization processing. It must also mention the origin of the product, as is the case for fruits and vegetables. Lastly, the use-by date must be visible, for example on the silo or tray, and must be updated as trays are refilled. For non-food items, there are specific regulations, in particular for cosmetics and detergents. The range of packaging-free products can therefore be extended through legal developments. Which products are currently sold without packaging and is this likely to change?

5.3.2.2. *Towards a progressive expansion of products on offer: support for innovation*

According to a Nielsen study in 2018[32], the most purchased food products sold unpackaged are oleaginous fruits, such as nuts, almonds and hazelnuts (bought by 58% of respondants), dry fruits (47%) and legumes, such as lentils and chickpeas (32%), and cereals (28%). For non-food items, 6% of respondants said that they buy their washing-up liquid unpackaged. O. Dauvers[33] indicates that, generally speaking, the range offered in sales outlets remains limited, even though there is a trend towards further development, whether in terms of food (e.g. oils, yoghurts) or non-food items (particularly cosmetic products). One of the restrictions on range is related to the preservation of food products at room temperature[34]. Space restrictions in stores can also be important, especially if the choice is made to offer packaging-free products and processed items for the same category of product together.

Some chains are starting to offer types of products that had not previously been sold packaging-free. For instance, the L'Eau Vive network of organic shops is offering

31 The DGCCRF (*Direction Générale de la Concurrence, de la Consommation et de la Repression des Fraudes*) is a French administrative body under the Ministry of the Economy. Its purpose is to oversee the conditions of trade between companies in order to ensure the proper functioning of markets for the benefit of consumers and companies. See: https://www.economie.gouv.fr/dgccrf/Publications/Vie-pratique/Fiches-pratiques/Etiquetage-des-denrees-alimentaires.

32 See: https://www.nielsen.com/fr/fr/insights/news/2019/bulk-buying-habit-that-settles-in-the-french.html.

33 Observatoire du rayon vrac 2019 [Online]. Available at: https://www.olivierdauvers.fr/2019/10/18/en-libre-telechargement-lobservatoire-du-rayon-vrac-2019/.

34 Faut-il s'emballer pour le vrac ? *LSA*, 2419, June 30, 2016.

filtered water as a packaging-free product. This has been made possible through the use of an innovative device and the availability of reusable hard plastic containers[35]. With regard to liquids, the packaging-free furnishings supplier Jean Bouteille also provides an example[36]. Three types of product are offered in these shops: oil, vinager and wine. The customer helps themselves using a fountain that automatically distributes the desired volume to match the chosen bottle. C. de Boisse (Briochin) also refers to the use of such fountains for cleaning products: "They make, for example, 50 cm large pieces of furniture that can hold four fountains that distribute several different products, such as cleaning products, detergent, washing-up liquid, to meet all of the expectations and needs of consumers in terms of cleaning/ washing." A packaging-free yoghurt dispenser was also installed in a Day by Day shop in 2019, in partnership with the Danone and Faire Bien brand of organic yoghurts. The aim was to meet consumer expectations, while looking to provide ultra-fresh packaging-free items.

These examples of innovation illustrate the need, for some types of products, to rethink the kind of devices offered in store and the merchandising to develop them.

5.3.3. *The "logistics, distribution and merchandising" triptych for packaging-free products*

Providing a product without packaging involves restrictions in terms of the organization of store floor space, cleaning, refilling and cleaning devices[37]. Proposing new categories of packaging-free products thus involves rethinking the "logistics, distribution and merchandising" triptych.

5.3.3.1. *Specific logistics for packaging-free consumption*

From a logistical perspective, the question of procurement arises from suppliers and, in particular, the desire to favor local suppliers to follow a more responsible approach. With regard to storage, it is also important to have a big enough storage space, as packaging-free products require bags to be stored once they have been opened. The Biocoop La Ruche de Tolbiac store has, for example, put in place special tongs for closing packets and has a file to save the labels from each bag and record the date they were each opened to ensure proper traceability[38]. Refilling and

35 L'Eau Vive vend de l'eau en vrac. *LSA*, special issue, March 2019.

36 Jean Bouteille inaugure le vrac des liquides. *LSA*, 2314, April 2014.

37 Faut-il s'emballer pour le vrac ? *LSA*, 2419, June 2016.

38 Meubles de vrac : la méthode Biocoop. *LSA*, 2536, December 2018.

cleaning devices occurs at various levels, in stores or on dedicated platforms. Day by Day, for example, has opted for a comprehensive platform split into three spaces: an industrialized washing room for before the trays are filled; a room to prepare and fill trays; and a storage area to separate organic and non-organic products. At Jean Bouteille, bottles are recorded and cleaned in a cleaning unit attached to the store, which limits logistical costs. Logistical issues will thus go hand in hand with general issues of hygiene. In store, the regular cleaning of equipment is necessary to avoid the proliferation of insects and fungus, particularly for liquids (e.g. juice, syrups, soups) and for ultra-fresh items (e.g. yoghurts, ice creams)[39]. This cleaning requires human resources and innovation, following the example of the small or medium-sized enterprise Ecobox, which proposed mobile containers that would be easy to clean. It should be noted that the consumer perception of hygiene is generally better for silos, where products enter from the top and leave from the bottom, than for containers where the products enter and exit through the same channel.

5.3.3.2. *The evolution of distribution and merchandising*

In terms of distribution and merchandising in store, selling packaging free products requires changing the furnishings, reconsidering arrangements and ensuring proper storage conditions for products (temperature and heating, for example). The distribution of packaging-free products occurs primarily through silos or trays, but other innovations are being developed. The new Zero Waste space of Franprix Réaumur, for instance, is equipped with cardboard funnels[40]. We can also mention the "DoZeuse" innovation, an award-winner in 2019 and used by the CoZie brand, which offers organic *made in France* products in reusable and returnable glass bottles[41]. This patented system allows customers to serve themselves the right amount and respect hygiene rules, which are ever stricter for cosmetic products.

The issue of payment is also a challenge, as packaging-free items require the operation of scales to be reconsidered[42]. Some chains offer different scales depending on the type of product. Stores can also choose to use simple scales or connected devices to make it easier manage stocks and accounts. The availability of scales for customers, or lack thereof, and their placement are also issues to consider,

39 Tendance : en plein boom le vrac a encore du chemin à faire. *LSA*, September 2019.

40 Le nouveau Franprix pousse la restauration et le zéro déchet. *LSA*, 2572, September 2019.

41 See: https://www.linfodurable.fr/conso/zero-dechet-quand-les-cosmetiques-passent-au-vrac-7043.

42 Le vrac transforme l'équilibre des balances. *LSA*, 2551, April 2019.

as well as the labeling of products and of containers brought by customers to ease the packing and the customer's journey[43].

The sale of packaging-free products thus involves both improving the ergonomic design of the range to facilitate product identification, and easing the weighing stage by allowing customers, where necessary, to weigh their reusable containers. This therefore requires specific customer support.

While the retailers have to contend with the constraints affecting the development of packaging-free consumption (legal and logistical ones, in particular), they must also support customers in the unique packaging-free buying process.

5.3.4. *Information support for consumers of packaging-free products*

Faced with packaging-free products that can be considered new "OCNI" (*Objets comestibles non identifiés*, unidentified edible objects (Fischler 1990)), consumers must relearn the rules of consumption. Retailers must therefore ask themselves how to disseminate relevant information. In order to ease the "task" for their customers, chains must support them and, in particular, provide them with all of the necessary information. This can be delivered by various means, such as labeling and information in the sales location, as well as through sellers present in store.

5.3.4.1. *Price information: a key point*

As highlighted earlier, price can be a determining factor in the decision of some consumers to choose packaging-free products. They want to have an idea of the price of their chosen quantity and assure themselves that they have got a "good deal" in relation to the equivalent packaged product. For the more reluctant, it would appear that the lower price inherent in packaging-free consumption is not always proof, with some consumers sometimes noticing significant differences in price between a packaging-free product and its packaged equivalent, not always in favor of the packaging-free product, and then denouncing it on social media[44]. It should be noted that, in such cases, the comparison may involve an organic product and a non-organic product.

---

43 White paper. Gérer un espace de distribution vrac [Online]. Available at: http://vracoop.fr/guide-du-vrac/.

44 See: https://www.capital.fr/lifestyle/les-produits-en-vrac-ne-sont-pas-forcement-moins-chers-1337299.

However, it is necessary to compare comparable products. Retailers thus try to support consumers in making price comparisons to reassure them that packaging-free products are indeed between 5 and 30% cheaper than their equivalent packaged products[45]. They thereby emphasize the price of a shopping cart with packaged products and the price of another with comparable packaging-free items. Labels can also be placed on some products to inform consumers of low prices.

Moreover, consumers can sometimes feel lost when confronted with prices indicated on silos and trays. In fact, the prices of packaging-free products are often set by weight, either by the kilo, generally, or per 100 grams. Consumers can therefore find it difficult to find the price when the reference weights indicated vary and the comparison between a packaging-free product and its packaged equivalent is not straightforward.

5.3.4.2. *Necessary customer education?*

Customers can quickly find themselves inept or incompetent when buying methods change and, specifically, when there is no longer any packaging (Fuentes *et al.* 2019). Indeed, throughout the packaging-free buying process, consumers need to be supported and educated in: the choice of products (finding products in a range of silos, understanding prices, recognizing unfamiliar products, assessing their quality, etc.); their mode of preparation (cooking time, for example); how to handle them (self-picking in trays and silos, quantity); the use of equipment and scales (weighing, scanning product codes); and even product transport (bringing containers).

From the perspective of the development of packaging-free consumption, customer education is essential[46] and relearning the useful terms for making purchases seems necessary. Since the late 1980s, driven in particular by technological innovations, self-service devices have been put in place (Mencarelli and Rivière 2014), such as weighing, self-scanning and self-packing. This is also the case for sales of packaging-free products. These measures co-exist with salaried workers, necessary for organizing products and for customer support. In chains, there is thus important "work" that goes on between customers and employees who communicate on the shop floor and at the cash regiter. With the development of sales of packaging-free products, this will likely develop further, thereby redefining the role of sellers, who customers must (re)learn to talk to.

45 Faut-il s'emballer pour le vrac ? *LSA*, 2419, June 30, 2016.

46 Meubles de vrac : la méthode Biocoop. *LSA*, 2536, December 20, 2018.

5.3.5. *Revisiting the role of the seller*

In the sale of packaging-free products, the role of the seller in relation to the customer is key. The seller must at once advise customers with their choice, support them during the buying process and also monitor them to avoid any poor handling that could create losses. In addition, their role is changing and involves loading, maintenance and cleaning to ensure that the store looks its best: in the absence of packaging, the presence of a seller is of the utmost importance. "The shop must be spotless to instill confidence in its hygiene," confirms Nathalie Garde (manager of a Day by Day shop). *In the seller's store, there is neither a grain of rice nor a drop of detergent to be seen. The seller must be available at all times to help to operate the pasta distributor, help a customer to decide on a ground coffee or provide information on a given brand of cleaning product*[47]. This feeds into the creation of a trust-based relationship: "the seller occupies a key role in the grocery store. Consumers expect their advice and support whatever the type of sales, but this is even truer of a grocery store almost entirely stocked with packaging-free products"[48]. The seller of packaging-free products thus assumes several roles.

The role of advisor: similar to open-air markets, buying packaging-free products does not only involve face-to-face contact between a buyer and a product. It also involves sellers, whom the customers ask about, for instance, the origin of the products, their quality, or advice on how to prepare them. This therefore involves moving beyond strict sales towards a kind of pedagogy. M. Cadic (Mamie Mesure) explains:

> To replace the silent seller, it is our expertise, our advice, our availability that matters. We always try to approach people without disturbing them, all the while saying "well, don't hesitate to ask if you have any questions, if you don't understand what a product is for…" There is an awful lot of teaching involved. When recruiting someone in stores like ours, it won't be product knowledge, but rather personal conviction, their own personal zero waste lifestyle, that will be more important than sales competence, or specific courses they would have taken.

The support role: in packaging-free consumption, the seller can play an important role in informing the consumer about products, as well as about the "rules" of packaging-free consumption (e.g. bringing your own containers,

47 See: https://www.lemonde.fr/m-perso/article/2015/12/14/la-vente-en-vrac-seduit-les-consommateurs_4831637_4497916.html.

48 See: https://www.epiciersdefrance.org/blog-epiciers/vrac-retour-aux-sources.

weighing). Chains must thus find solutions on how to provide the information traditionally included on packaging (e.g. cooking time, recipes, nutritional information)[49]. M. Cadic emphasizes:

> As we're a small shop, we're used to it. We show how it works. In any event, there is the obligatory stage [of weighing products] at the scales before self-service. So we ask customers "have you already served yourself? Do you know how this works? Do you need help?" And we prefer support. Without it, for us there is the risk of product loss and a lot of cleaning up after.

The monitoring role: in the context of packaging-free sales, one question can legitimately be asked. Are customers at risk of increasing "bad" behaviors, intentionally or otherwise? As Hocquelet (2013) indicates, by handling the products in store themselves, customers may create reductions (damaged or broken products) or change their mind and put items back, sometimes after they have been partially consumed. Forms of trickery, or even theft, can also occur (changing weight labels, for example). This contributes to an increased workload for employees and potential losses for the chain. It is therefore very important to monitor customers and, especially, to educate them about packaging-free consumption and create a close relationship in-store between sellers and consumers.

The role of guarantor of quality and cleanliness: the seller of packaging-free products must continuously and rigorously keep the products looking good. This well-kept appearance is a reflection of the quality of the products and the chain in the eyes of consumers. Moreover, the improper use of products and equipment can create damage (products inadvertently left in the sun, for example). Sellers must therefore ensure that other customers do not see this potential damage.

Owing to their multiplicity of roles, the seller of packaging-free products has become a means of instilling consumer trust in the chain, as is the case in open-air markets. Indeed, as highlighted by Aurier and Sirieix (2016), in the context of fresh products sold at markets, there is an attachment to the sellers more than to the products. The majority of buyers acquire their goods from a small number of sellers with whom they build a relationship based on trust. Beyond the role of sellers, the role of other customers can also be key in the education of new customers: children can thus participate in a "reverse socialization" process and teach their parents, or long-standing customers can teach newer ones.

49 Faut-il s'emballer pour le vrac ? *LSA*, 2419, June 30, 2016.

5.4. Conclusion

The packaging-free market is currently undergoing a double growth with, on the one hand, new consumers who would not previously have been drawn to this mode of shopping, and, on the other hand, more committed consumers who are no longer just a minority. The new consumers are very varied. Some choose packaging-free consumption for environmental reasons and others come to it because of the efforts made to diversify the products on offer, and remove obstacles to accessibility and practicality. D. Onraita explains that, for Day by Day, the two key elements are regional coverage outside big cities ("We currently cover 47 departments and it is essential that we cover all of them by the end of 2022, including Lozère") and showing that packaging-free consumption can be adopted "by everyone" without much effort ("We have to explain to everyone that this packaging-free mode of consumption makes sense, that it isn't sad, that it isn't backwards, and that it can, in fact, be adapted to the realities of daily life"). At the same time, the habits of those convinced by packaging-free consumption, who may have previously appeared to be few and far between, disorganized or part of a self-sufficiency movement, are starting to become more normalized. Individuals, such as Béa Johnson or the Zero Waste Family, who have driven the Zero Waste movement since its inception and are institutionalizing it, today embody an ideal for numerous consumers looking to simplify their modes of consumption.

Lastly, it is possible to question the ability of packaging-free consumption to participate in the rebirth of sustainable development. This chapter has shown the potential advantages of packaging-free consumption, but this market is still too new and too limited for there to be reliable data to confirm these potential advantages. Moreover, while packaging-free consumption is not a new mode of buying, can it be considered a solution to waste if practical daily issues (food storage, quantities cooked and consumed) are not addressed (Dyen *et al.* 2018)? More generally, questions remain about the manner in which packaging-free consumption could become part of a more comprehensive vision of sustainable development, including Zero Waste objectives and the circular economy at the level of individual households.

5.5. References

Argo, J.J., Dahl, D.W., Morales, A.C. (2006). Consumer contamination: How consumers react to products touched by others. *Journal of Marketing*, 70(2), 81–94.

Aurier, P. and Sirieix, L. (2016). *Le marketing de l'agro-alimentaire : environnement, stratégies et plans d'actions*. Dunod, Paris.

Barré, B. and Heilbrunn, B. (2012). *Le packaging*. Presses universitaires de France, Paris.

Beitzen-Heineke, E.F., Balta-Ozkan, N., Reefke, H. (2017). The prospects of zero-packaging grocery stores to improve the social and environmental impacts of the food supply chain. *Journal of Cleaner Production*, 140(3), 1528–1541.

Cochoy, F. (2002). *Sociologie du packaging ou l'âne de Buridan face au marché*. Presses universitaires de France, Paris.

Daniel, M. and Sirieix, L. (2012). Les pratiques durables : une forme de résistance ordinaire ? *Décisions Marketing*, 68(4), 11–24.

Dyen, M., Sirieix, L., Costa, S., Depezay, L., Castagna, E. (2018). Exploring the dynamics of food routines: A practice-based study to understand households' daily life. *European Journal of Marketing*, 52(12), 2544–2556.

Ferrières, M. (2002). *Histoire des peurs alimentaires. Du Moyen Âge à l'aube du XXème siècle*. Le Seuil, Paris.

Figuié, M. and Bricas, N. (2016). Faire ses courses au Vietnam aujourd'hui : quand les supermarchés touchent aux sens. *Anthropology of Food*, Articles VARIA.

Fischler, C. (1990). *L'homnivore*. Odile Jacob, Paris.

Fuentes, C., Enarsson, P., Kristoffersson, L. (2019). Unpacking package free shopping: Alternative retailing and the reinvention of the practice of shopping. *Journal of Retailing and Consumer Services*, 50, 258–265.

Hocquelet, M. (2013). L'expérience des salariés des hypermarchés. *Revue française de gestion*, 5(234), 135–149.

Mencarelli, R. and Rivière, A. (2014). La participation du client dans un contexte de self-service technologies : une approche par la valeur perçue. *Revue française de gestion*, 4(241), 13–30.

Monnot, E. and Reniou, F. (2012). Les suremballages : des emballages superflus pour les consommateurs ? *Décisions Marketing*, 65(1), 31–42.

Rozin, P. and Nemeroff, C. (2002). Sympathetic magical thinking: The contagion and similarity "heuristics". In *Heuristics and Biases: The Psychology of Intuitive Judgment*, Gilovich, T., Griffin, D., Kahneman, D. (eds). Cambridge University Press, Cambridge.

6

The Conditions for Effective Social Communication[1]

Agnès FRANÇOIS-LECOMPTE[1] and Sylvie FOUTREL[2]

[1] *LEGO (Laboratoire d'économie et de gestion de l'Ouest), University of South Brittany, Vannes, France*

[2] *LEGO (Laboratoire d'économie et de gestion de l'Ouest), Catholic University of West Brittany South, Arradon, France*

6.1. Introduction

While the concept of corporate social responsibility (CSR) is not new, it seems to have become more and more widespread within the economic fabric in recent years. It refers to the voluntary integration by companies of social and environmental concerns into their commercial activities and their relationships with stakeholders (European Commission Green Paper – July 18, 2001). While CSR refers to a theoretical field that is older and applied more to the business world than sustainable development, the two notions are, in fact, currently used interchangeably without any real distinction (Mauléon and Silva 2009). Within this study, CSR is therefore assimilated into the act of companies taking sustainable development into account. The extensive study carried out by EcoVadis (2019) on 21,000 companies worldwide indicates that the integration of social and environmental criteria has increased year on year: in France, 70% of small and medium-sized enterprises and 75% of large enterprises now have a CSR management system. This movement is partially explained by a regulatory context that obliges large companies to confront

1 The authors wish to warmly thank the Sustainable Development Manager of the Triskalia group for her participation in this chapter.

Marketing for Sustainable Development,
coordinated by Sihem DEKHILI. © ISTE Ltd 2021.

these challenges and take them into account by means of an extra-financial relationship (Grenelle II Law). Moreover, there are high social expectations that businesses will assume their share of responsibility in the green transition. According to a Greenflex survey, 61% of consumers expect companies to act in support of sustainable development (Greenflex 2019). In addition, the benefits of CSR on economic and commercial performance, the human resources environment within the company and even the risks to reputation are no longer in question.

That being said, the question to be addressed here concerns the value of this type of commitment to company stakeholders and, more specifically, to customers. While the concept of sustainable development is popular among the general public, the links between marketing and CSR are complicated and riddled with paradoxes.

While consumers expect companies to engage, the reactions to social communications are often negative: one survey (*Le regard des Français et leur appropriation de la RSE* – The French view and their appropriation of CSR, (Ifop 2019)) shows, for example, as many positive as negative opinions on the subject. Discussion of CSR is at the heart of two opposing movements: the legitimate desire of a company to make its commitments known to the public, on the one hand, and the risk of being accused of opportunism and "greenwashing" on the other. The exercise seems so risky that many companies prefer to stay quiet about their actions in support of sustainable development (Delmas and Colgan 2018). However, some companies have demonstrated that it is possible to make the case for CSR to customers, and to even make it central to their brand. Unilever's social shift is a good example. Sustainable marketing can be a means of creating a better brand image and, more simply, of increasing market share (White Paper 2016). The question raised in this chapter is therefore how a company can profit from its activities in support of society and the environment in the marketing of its offer. In other words: How can CSR be promoted among its customers?

The question will be addressed here on the basis of a review of the academic and professional literature on the topic. The aim is to create a set of recommendations for companies from academic and professional experts in the field. The ideas will be grouped into three sections: the presentation of the framework used to address the question (section 6.2), advice on ensuring the credibility of CSR communication (section 6.3) and recommendations for making CSR an added value for customers (section 6.4). The desire to compare words with reality has led us to gather testimony from the Triskalia group. This is presented at the end of the chapter (Box 6.1), with the questions asked of its Sustainable Development Manager being reflected in the statements made in the three sections of this chapter.

6.2. Social communication: a shifting reality

Companies can follow different CSR communications strategies. The aim of this section is to examine the various potential options managers can take and the challenges they entail. This also involves consideration of how these strategies can be put into operation.

6.2.1. *To say or not to say?*

While CSR refers to one reality, CSR communication refers to another. The actions of a company in society or the environment (the "doing") do not necessarily mean that they are communicated to stakeholders (the "saying"). Social communication therefore refers to "all communication activities, regardless of the medium, that deliver a message about the environmental, social or societal commitments of an organization" (Benoit-Moreau *et al.* 2010). In this sense, societal communication can be seen as the tip of the iceberg (Grant 2009). However, it is only through this that stakeholders can learn about a company's social or environmental efforts.

There are many possible strategies for social communication; before addressing the crucial matter of content, the equally crucial matter of intensity rears its head. To return to the metaphor of the iceberg, the first question to be asked relates to the height of the tip of the iceberg. The visible part of CSR varies between companies and the challenge is to avoid two pitfalls: not profiting from CSR in its business relationships or, conversely, saying too much about the company's activities in this area.

Figure 6.1 presents the three schematic positions of intensity in social communication.

The view defended in this chapter is the median position in Figure 6.1, where CSR is visible to a company's stakeholders but not to an excessive degree. The two other positions present a number of risks that we examine.

The position on the left (Figure 6.1) refers to the case of a company working in favor of sustainable development but without informing its stakeholders of this or, otherwise, being satisfied with mandatory institutional communication on the subject[2]. This first situation appears to be quite common (Delmas and Colgan 2018).

---

2 In France, companies with more than 500 employees are obliged to take into account the social and environmental impact of their activities in an annual document, known as a non-financial

One of the reasons companies stay quiet about CSR is the desire to avoid the inherent risks of such a practice. A responsible marketing approach, i.e. integrating CSR into the marketing strategy, can flounder and result in a loss of business or market share (Jahnich 2013). It is natural to act cautiously and ensure a return on investment by making a strategic choice. As well as this economic risk, there is a risk to reputation and image. Consumers are often suspicious when a brand refers to its commitment and may see it as an opportunistic attempt to make money (Robert-Demontrond and Joyeau 2010). Communicating therefore means creating expectations among stakeholders and taking the risk of being accused of using sustainable development for commercial purposes (Benoit-Moreau *et al.* 2010). There is also the risk of its communication being discussed, denouced or attacked by stakeholders outside the company (journalists, associations, citizens, etc.) (Du *et al.* 2010). There can be a strong temptation to remain discreet regarding social engagement (van de Ven 2008), but this strategy has the downside of not being able to profit from the CSR approach in commercial terms. However, sustainable development can create value for a company's customers (Gabriel 2003), and act as a means of differentiating a brand (Laville and Florentin 2015) or a growth lever for the company (white paper 2016). Not linking marketing to CSR (see Figure 6.1) is therefore a choice commonly made by companies, thereby minimizing risks but also leading to the company missing out on important opportunities.

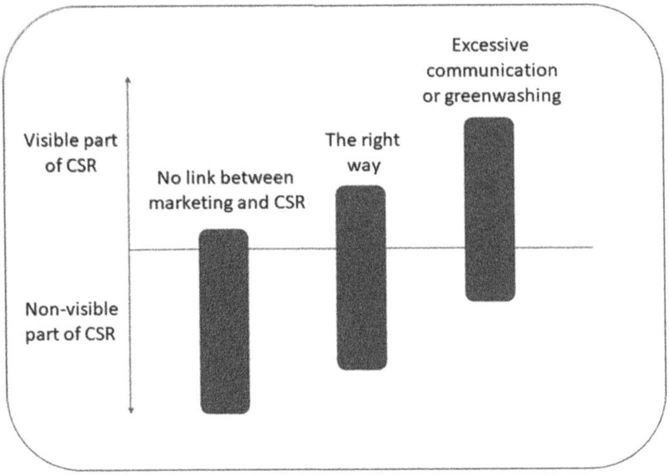

Figure 6.1. *Overview of potential social communication strategies. For a color version of this figure, see www.iste.co.uk/dekhili/sustainable.zip*

performance statement or sustainable development report (Grenelle Law, 2010; European Directive on non-financial reporting 2017).

The position on the right in Figure 6.1 refers to an opposite situation where the company communicates its commitment to sustainable development to an excessive degree. This refers to the concept of "greenwashing" which is the abusive or misleading use of environmental arguments in advertising. In France, this practice is regulated and monitored by the *Autorité de régulation professionnelle de la publicité*, ARPP (Authority for the Professional Regulation of Advertising) and its subsidiary bodies. While there seem to be relatively few cases of greenwashing according to the statements issued on the subject by the *Agence de l'environnement et de la maitrise de l'énergie*, ADEME (French Environment and Energy Management Agency) and the ARPP[3], the practice has unclear boundaries. It is thus associated more broadly with all situations where companies communicate loud and clear about sustainable development even when their environmental performance is poor (Delmas and Burbano 2011). For example, this is, the case for companies that spend more money on their advertising budget to highlight their support for a cause than on the actual cause itself (Grant 2009). Such cases are obviously not to be encouraged as they mislead consumers and feed into the skepticism of the general public on these issues. Such practices are monitored and quickly denounced by numerous bodies, putting the reputations of companies communicating in a misleading manner at risk. Credibility is hard-won, making greenwashing ultimately much more costly than it initially appears (Grant 2009).

The central position in Figure 6.1 refers to the case where the company transforms its CSR into added value for stakeholders, without communicating it too much. This is therefore the case that we develop in the rest of this chapter.

6.2.2. *A triptych to be adapted to different situations*

Social communication covers various types of company practices, ranging from institutional communications, such as reports on sustainable development, to advertising, to information on packaging. It is useful to establish some frameworks for studying this concept. Broadly speaking, social communication materializes through choice on three main aspects that are adapted to the situation of the business (Du *et al.* 2010): the target, the channel and the content.

With respect to the target, social communications are potentially addressing all of the company's stakeholders: public authorities, the press, non-governmental organizations (NGOs), investors, employees, the local community, consumers,

3 See *Publicité et environnement*, (ARPP/ADEME 2017). Available at: https://www.arpp.org/wp-content/uploads/2016/07/Bilan_DD_2015.pdf.

retailers, etc. (Du *et al.* 2010). Very often, there are one or several main targets and the communication format must be adapted accordingly. A sustainable development report is thus addressed to the institutional partners that the company must take into account, while advertising is addressed primarily to customers, for example. Customers, the population studied here, are often not the main target for social communications. For a cooperative such as Triskalia, members are often a priority target.

Secondly, social communication takes place through channels. These can include advertising, official documents such as sustainable development reports, websites, promotional means, the sponsoring of events dedicated to a social cause, public relations, product packaging and sales initiatives. There are numerous means and they are suited to different aims. It is important to note that, among all of these channels, some are used by the company while others are beyond its control (Du *et al.* 2010). These external channels, such as the media or consumer forums/blogs, echo the company's messages and can sometimes send contradictory signals (Benoit-Moreau *et al.* 2010).

Lastly, the third pillar of social communication is its content. On this point, as shown in Figure 6.1, the business often makes a choice in what it decides to make visible to its stakeholders and the subjects it does not wish to mention. CSR often leads to a range of actions relating to eco-design, HR strategy, sustainable purchasing, sponsorship, the energy efficiency of buildings and so on. It is therefore necessary to establish priority elements to be highlighted to customers (Schlegelmilch and Pollach 2005).

In sum, social communication can be seen as a triptych on the questions of "what message", "which channel" and "to whom". The choices must naturally take into account the specific situation of each company, including its reputation, its competitive context and the potential positioning of its brands in relation to CSR (Du *et al.* 2010). The next section of this chapter provides the recommended elements of this decision-making triptych by taking a transversal approach. The academic and professional literature in fact shows two major challenges that are addressed one after the other: the credibility of social communication (section 6.3) and the need to provide the client with added value (section 6.4).

6.3. How can the credibility of communication be ensured?

As mentioned, the current context is one of distrust when companies take on the subject of sustainable development. While consumers expect companies to engage, many of them react negatively to company actions being highlighted.

The denunciation relates largely to the sincerity of the approach and the accuracy of statements (Robert-Demontrond and Joyeau 2010). NGOs are quick to accuse communication efforts of greenwashing, and not necessarily fairly (Laville and Florentin 2015). In short, the success of a social communication campaign relies in large part on the credibility of the message (Schlegelmilch and Pollach 2005). In other words, it is a matter of being perceived by consumers as trustworthy and an expert on the subject (Newell and Goldsmith 2001). To that end, a number of precautions need to be taken.

6.3.1. *Communicating using proof*

An essential recommendation for companies wishing to be taken seriously relates to the communication of concrete actions rather than vague social or environmental claims (Gabriel 2003). This therefore involves providing proof of the company approach as the consumer may otherwise doubt the message. For example, this can, involve providing pledges such as specific and quantified assertions. One of the "Sustainable Development" recommendations of the ARPP refers to this principle: "The speaker must be able to justify their arguments referring to sustainable development with serious, objective and verifiable elements at the time of advertising." The trend thus seems to be towards presenting approaches by means of quantifying the efforts made or the effects in terms of their social and environmental impact (Du *et al.* 2010). One of the typical techniques based on this logic is product sharing, where a company commits to investing a given amount of money in a good cause for each consumer purchase. The brand Pampers, for example, committed to providing a vaccination to UNICEF for each pack of diapers bought in the last three months of 2006. The artisanal Brittany brand Les Surgelés (Triskalia group, see Box 6.1) provides a contribution to farmers coverting to organic agriculture for each frozen food item purchased. In another vein, while eco-design approaches are being developed, it is advised that the amount of raw materials saved be specified. The B2P pen from Pilot thus claims to be made of 89% recycled plastic. Conversely, any vague assertion of being "better for the environment" or a "green product" or similar will be more easily questioned by consumers – particularly by those who are better-informed – and denounced as greenwashing.

It should be noted that, in France, regulations now require large companies to publish an annual statement on their non-financial performance with a certain number of numerical indicators: the logic of quantifying environmental and social performance thus seems to be becoming more widespread.

6.3.2. *Seeking out external guarantees*

A second recommendation for lending credibility to communications relates to external guarantees. The general idea is that the social or environmental approach will have a better chance of being taken seriously if it is guaranteed by or carried out in partnership with a third party. In other words, "the legitimacy of the company cannot be self-proclaimed; it can only be acknowledged" (Gabriel 2003, p. 72). Partnerships thereby aim to make use of another established and recognized "brand" in the field where the company wishes to be visible and credible (Grant 2009). Certifications and product-sharing operations are the two main ways of benefiting from the reputation of a third party (Grant 2009).

In the first case, there is a large variety of labels that can be attached to products or services. Generally speaking, they bring a guarantee of respect for social and environmental causes to a range of products or the company as a whole. The requirements and demands of the certification process depend on labels.

In terms of the environment, Organic Agriculture and the European Ecolabel are among the most widespread public labels. In social terms, the Fairtrade Max Havelaar label, although private, was one of the first to provide guarantees of the working conditions of its producers. Over time, the list of labels and certifications available to companies has continually expanded, offering a multitude of specialized tools for a given subject (for example, organic, fairtrade) or a guarantee of a company's global approach (for example, ISO 14001, AFAQ 26000). While the labeling approach is often expensive, it generally offers a return on investment whether in a reduction of costs or in commercial visibility. It is also a way of federating the approach and acquiring a set of specifications to follow. Labels, in particular those provided by a third-party institution, increase consumer confidence (Dekhili and Achabou 2013). However, the effect is not systematic and the impact on sales depends on numerous other parameters. On the other hand, CSR certifications and labels are gaining importance in the "Business to Business" context and in public tenders (Sustainable Procurement Barometer 2019). Indeed, they constitute a means of tangible assessment for companies implementing a sustainable procurement policy.

In the case of product-sharing operations, the brand is associated with public interest for a win-win situation (Varadarajan and Menon 1988): NGOs benefit from an additional budget while the reputation of the NGO reflects back positively on the company.

Despite this, these operations are criticized as they make people believe that their actions are more effective than they really are, and on the grounds that the company sometimes spends more money on making their actions known than on the cause supported (Grant 2009). Nonetheless, it is a formula that has proven its worth and makes it possible for companies that prefer not to mention their internal practices to launch a sustainable marketing approach.

6.3.3. *Getting others to talk about you*

Another major aspect of credibility relates to the communication channel used to transmit the information to consumers. On the one hand, there are means that the company controls (website, packaging, reports, etc.) and, on the other hand, those that it does not (media including social networks, forums/blogs) because they come from sources outside the company, such as NGOs, opinion leaders, consumers and journalists. Paradoxically, the more a channel can be controlled, the less credible it is, and vice versa (Du *et al.* 2010). Advertising, for example, is a channel poorly suited to communicating sustainable development actions (Schlegelmilch and Pollach 2005). In reality, consumers form an opinion about a company on the basis of the views of its stakeholders (Morsing *et al.* 2008). The ideal but difficult to achieve scenario is for independent sources to promote the company. "Today, when we have the best products and the best policies, it is NGOs, the CSR forum, the consumer ethics guide or the journalist that takes care of your advertising" (Grant 2009, p. 74). This requires media coverage (TV, press, Internet), or, in other words, the creation of positive buzz around the company. In order to do this, it is important to provide plenty of information about the company, for example on its website, to respond to requests for interviews, to take original or significant initiatives, to be visible during CSR events or even for the company to put themselves forward for awards or prizes relating to social or environmental issues. With the rise of social networks, consumers are becoming increasingly important in providing word-of-mouth information on company practices (Du *et al.* 2010). It is therefore in the company's interest to anticipate and try to guide consumer reactions to their CSR. Storytelling based on accounts of CSR commitments is one example of a means of creating emotion among the general public.

In addition to sources outside the company, employees remain important actors in passing on information about the commitments of their company. When they are convinced of the sincerity of the approach, they can become real ambassadors by speaking to their family and friends. It is therefore important for a company to aim at this target group and to see how they can be used to relay CSR information to the general public (Du *et al.* 2010).

6.3.4. *A long-term commitment*

Lastly, social communication can appear more sincere in the eyes of the public when the company is not new. The more history of CSR the company has, the more legitimate it will be to talk about it publicly (Du *et al.* 2010). On the other hand, a timely communications campaign on a subject is easily seen as a self-interested approach. The message should thus preferably concern the social or environmental subjects that the company is familiar with and with which it has engaged over the long term. When it comes to environmental communication, it is therefore worth taking some time to consolidate one's commitments before including them in a marketing campaign.

In this regard, Morsing *et al.* (2008) advise raising awareness of CSR internally before communicating it externally. The first stage therefore involves ensuring a certain level of visibility and recognition among employees. Only then in a second stage should the company gradually begin to make its commitments known to external stakeholders: first the local community, then a national and eventually international audience. It is therefore preferable to adopt a progressive approach where the company gradually expands the scope of its social communications.

6.4. How can CSR provide added value to customers?

A second key challenge in social communication relates to the creation of value for customers using CSR. Linking marketing and CSR in fact relies on CSR being a selling point or, at least, an element that reinforces the appeal of the offer. In addition, as well as making the company's social and environmental commitments visible (Figure 6.1), the aim of social communication is to increase their value among consumers. Three recommendations have been developed around this idea.

6.4.1. *Choosing the adequate themes of communication*

First of all, the topics that an organization decides to highlight are important. This choice also falls within the general process of building an image and therefore has major strategic challenges. All CSR communication thus participates in the construction of the "perceived image" of the organization among different groups and the difficulties of managers therefore involve ensuring that that image suits the "desired image" of the company (Schlegelmich and Pollach 2005).

In reality, it is impossible to add value to all of the social issues on which a company acts. The selection of the visible part of the iceberg can be based on two families of criteria that do not necessarily overlap:

– examine the company's objective social performance on different issues and choose those on which the company has made the most progress. It is actually particularly risky to communicate on the issue of CSR where the company is lagging behind or is simply content to follow the norms in force. Communicating creates expectations among target groups (Benoit-Moreau *et al.* 2010). It is thus necessary to be prepared to respond to such allegations, regardless of the origin of the criticism. In this context, communication must focus on the subjects where the company feels able to explain itself and is justified in doing so. Moreover, it must be taken into account that stakeholders expect a certain level of consistency between the CSR issues highlighted by the company and the activities it carries out (Du *et al.* 2010). A food company thus seems more able to comment on animal welfare, and non-GM and pesticide-free farming than a textile company, which is more concerned by child labor, climate change and transportation, for example;

– give precedence to issues of interest to consumers and try to differentiate themselves from competitors. Taking an interest in the subjects that concern consumers is often the easiest approach to implement. In fact, marketing services closely follow market trends. In France, for example, a recent Ifop study (September 2019) showed that the most anticipated CSR areas were working conditions and environmental protection, far ahead of the issue of diversity. It is therefore important to talk about the subjects anticipated by the general public, taking into account the fact that they vary over time and according to the news. For example, several years ago there was enthusiasm for products that were local. In order to grab consumer attention, it is also recommended that companies find a topic where they can differentiate themselves from their competitors and be perceived as unique (Schlegelmich and Pollach 2005). This approach can lead to the launch of new topics "outside" the company's main sphere of activity. This choice should be explained and justified to ensure that consumers do not disregard it and so that it can become consistent with the company's activity and history. In this marketing-based logic, talking in terms of CSR should therefore aim to create market differentiation or else be abandoned (Schlegelmich and Pollach 2005).

6.4.2. *Translating social engagement into customer benefit*

The trap that a company wishing to communicate its CSR approach could fall into would be to simply list the different criteria established in terms of its remit, standards, labels and certifications, to make an inventory of sorts and then publicize it.

Even if these direct encouragements have often made it possible to launch a process of reflection within the company alongside its legal obligations (CSR report, statement of non-financial performance) or, under pressure from consumers, for the company to engage in a CSR process, it is not just a question of ticking boxes because, ultimately, that does not motivate anyone either internally or externally.

For the consumer, what is important is for the company to convert its social commitments into client benefits and, at the risk of being satisfied by a niche market, not only for customers who are already convinced (Ottman *et al.* 2006). Responsible offers often have secondary advantages (they are more efficient, longer lasting, less expensive, better for health, etc.) and it is these characteristics that make a good sales pitch (Grant 2009). It is therefore important to understand what constitutes an asset for the consumer of a sustainable product or service and to emphasize the secondary benefits (Delmas and Colgan 2018). This reasoning has been particularly highlighted for green products.

The list of client benefits that can be linked to a responsible offers is presented in Table 6.1.

Secondary benefits	Examples of products
Quality/performance	Repairable equipment with a longer lifespan (Fairphone, Seb) Food products of a higher quality and with a better taste (organic products)
Price	Household appliances with an A++ energy efficiency rating Housing with an A energy efficiency rating
Health	Bébé Confort (eco-designed bottle with no products toxic to the baby) Organic products
Emotion	Traveling closer to nature (sustainable tourism) Feeling closer to producers (local products)
Status	Electric cars such as Prius Chaussures Veja (ethical sneakers worn by Marion Cotillard and Emma Watson, among others)

Table 6.1. *List of the main benefits of a responsible offer (source: created by the authors on the basis of Ottman* et al. *(2005) Delmas and Colgan (2018))*

Taking hold of all of these expectations to distribute environmentally-responsible products and practices could transform business to create a more sustainable world (Ottman *et al.* 2006; Delmas and Colgan 2018). In the absence of secondary benefits, it is also possible to use regular promotional means (price reductions, free samples, etc.) to increase the commercial appeal of responsible offerings and expand clientele beyond committed consumers (Ottman *et al.* 2006).

6.4.3. *Choosing the right tone for communications*

Beyond the information conveyed by the company, the thing that really matters is the relationship of trust established between the company and its customers. This relationship is built over time through the various interactions between a brand and its customers (advertising, exchanges on social networks, events, etc.). Here we propose four general pieces of advice – which should be adapted according to the media used – that make it possible to guide communications towards a more suitable tone:

– avoid anxiety-inducing, finger-pointing advertising: the counterproductive effect of this type of advertising has been proven and can even lead to negative engagement with environmental discussions (Monnot and Reniou 2013). Putting consumers face-to-face with problems far beyond them causes a kind of weariness, or even evasion. Some images, such as a polar bear lost and alone on an ice floe, have certainly made it possible to raise awareness of climate change but do not provide any concrete solutions on how to avoid it;

– use storytelling: as highlighted by Delmas and Colgan (2018), emotion is an important means of adherence to socially responsible approaches. The company can thus make use of storytelling techniques to construct its communications, especially for short formats on social networks. This is a matter of visualizing the approach, getting people to adopt it, talking about company commitments in the form of a story and, more broadly, revealing to the outside world what is happening within. As with the other approaches, the company must remain humble to avoid any accusation of greenwashing;

– do not "harm" the consumer experience: the purchasing behavior of a consumer, even an engaged one, is rarely guided solely by factors related to sustainable development. It is therefore necessary to use unobtrusive communications and not hide the main functions of the product. One of the challenges lies in pitching the sustainable message conveyed in the communication; it must be discreet enough to avoid disturbing the broader consumer experience, but comprehensive enough for responsible consumers who are most engaged. One possible solution consists of proposing different levels of reading social information adapted to different expectations and different target groups: for example, simplified on packaging but complete on a medium such as an app or a website;

– do not be afraid of using humor: lastly, it is recommended that companies favor positive, exciting and joyful communications, rather than lecturing (Ottman *et al.* 2006; Grant 2009). as The brand Le slip français has, for example, revolutionized the codes of communication around products made in France by leaning into provocation, self-deprecation and humor. Its success shows that the seriousness of the subject does not prevent the use of a light touch. Winning the consumer's attention and loyalty can therefore also involve amusement and pleasure, in particular at a time when social networks have demystified the company–consumer relationship.

Triskalia: cooperative food and agrifood group in Brittany, 22 brands (Paysan breton, Mamie Nova, Regilait, Ronsard, Magasin vert, etc.), 215 shops, 1.9 billion euro turnover, 4,800 employees. Triskalia merged with the d'Aucy Group in 2020 to form Eureden.

What do you make visible to your consumer customers, particularly in terms of your actions towards sustainable development?

That really depends on consumer expectations. We answer the questions they ask us and that's what we make visible, as well as ensuring our messages are well targeted. For example, for Ronsard's *Planète positive* chicken, we use terms like "no antibiotics", "slower growing", etc. These product features are the result of consumer concerns and our capacity to respond to them with our farmer members according to specifications.

On that point, how do you know what consumers think and what they want?

Our marketing and R&D people monitor the markets, trends ... There are also requests from our customers (retailers, manufacturers, catering) and our job is to respond to them [...] The direct contact that we have with the consumer comes from marketing studies, brand social media and sales events in store, such as by the pig farmers' Red Label.

What are your target groups when you communicate about your CSR?

If we take the example of our CSR report (or the statement on non-financial performance), there are different targets: employees and future employees aware of our CSR values, customers in general, institutional clients, etc. Our customers will be interested in the fact that the group as a whole respects values and carries out actions. This goes beyond the production of chicken or vegetables. It is about showing that today we are consistent in all that we do: a company that joins in with sustainable development cannot be excellent in one area and make a mockery of all the others!

Do you have links with marketing teams?

Yes, as marketers are our interface with customers and they need arguments and tools to be able to respond to their customers. What's more, they are also integrated into our CSR considerations. […]. Public procurement also increases their concerns. A non-financial activity statement allows us to respond broadly as the questions asked are often similar.

Why not make all of your approaches visible?

The first problem lies in the lack of time to do everything as we have a lot of activities and a lot of ideas! You have to be able to prioritize because not everything can be done at once. This is the whole point of analyzing our challenges: we prioritize our actions and our communications on our main challenges.

We haven't talked about them, but these anti-phyto stories affect you... how do you handle this in your communications plan?

When we are targeted directly, we manage our communications in response as best we can. So, in general, we try to show everything that we're doing well in terms of personal security and environmental protection. All communications come with risks because they can cause a backlash, but it has to be done anyway. We're not perfect, but we're taking action and making progress. For example, if we return to the issue of phytosanitary products, we have launched an action to reduce their use. We know that Rome wasn't built in a day, but we have communicated the fact that we launched this process all the same.

You have to be sincere to avoid greenwashing. I think that what's changing in company communications in general is that companies are prepared to say that they haven't yet achieved their objectives […] "we had a target of 50 and we're at 40, but we're making progress". There is this element of humility that is developing and I find that interesting, that no one's perfect and society can understand that!

In terms of other agrifood businesses, are you trying to find a very specific position?

Personally, no… I look at the other reports, but I don't think it helps us to identify our challenges because those are specific to us. However, we often come across the same topics, which isn't surprising. If we look at an agrifood product, the messages emphasized respond to the concerns of customers and the end consumer, and so they revolve around the same subjects [...]. In terms of institutional communication, we don't compare ourselves to others to make ourselves stand out, we just try to communicate in a way that truly reflects our challenges and our values.

Box 6.1. *Communications about sustainable development: interview with the Sustainable Development and Public Relations Manager of Triskalia*

6.5. Conclusion

While, at first glance, marketing and sustainable development appear to be pursuing different objectives, they can also be combined and push each other towards greater heights (Jahnich 2013). This chapter presents a non-exhaustive summary of the way in which marketing can support CSR in the commercialization of an offer. After a quick overview of the subject, a range of recommendations were proposed for integrated social communication. The testimonial from the Triskalia group, a major cooperative actor in the food sector in Brittany and one involved in sustainable development since 2011, illustrates a number of our statements. Firstly, only some of their engagements are visible to the general public as there are too many of them to all be conveyed to consumers. Mentions of social engagement seem ultimately to be quite discreet in relation to what could be communicated. This caution seems necessary. In terms of stakeholders, members and retailers' customers are the priority groups, especially as the group produces a lot for distributor brands. In order to establish credibility, communications are based on facts and supported by branding and labeling. Lastly, the subjects addressed are those of interest to the customers and that can help them to find personal benefits in relation to the quality of products and health.

The case of Triskalia also identifies a direct link with the subject developed: the connection between the marketing service and the sustainable development service. Quite often, it is the latter that is responsible for institutional communications, while the former is responsible for commercial communications. While links exist, they are not always facilitated in their day-to-day activities by the governance of the body. The challenge of sustainable marketing therefore consists of removing barriers between these two services and, more broadly, between the sustainable development service and the rest of the company. Moreover, social communication to the general public is not without risks and involves medium- and long-term strategic decision-making. Beyond links within the company, this involves a commitment and direction on the part of the company. Sustainable marketing therefore remains an emerging approach and one that requires a shift in the view of sustainable development: from an issue of compliance and the prevention of social damage to one of an integrated and differentiated vision of the markets (Laville and Florentin 2015). While it seems that few companies have turned this corner, CSR is increasing in importance in companies year after year, a trend that should only continue in the future. As mentioned, the process of translating this into commercial relationships brings many potential opportunities but also risks. There is no doubt that beyond the recommendations presented here there are many different potential paths towards taking on this challenge. A new generation of entrepreneurs, aware of the climate challenges and equipped to meet them, have been making themselves known in recent years.

A number of them have made CSR, and responding to social challenges more broadly, central to their projects. By continually connecting their mission statement to sustainable development, they embody a new way of doing business.

6.6. References

Benoît-Moreau, F., Larceneux, F., Parguel, B. (2010). La communication sociétale : entre opportunités et risques d'opportunisme. *Décisions Marketing*, 59, 75–78.

Dekhili, S. and Achabou, M. (2013). L'efficacité des labels environnementaux auprès des consommateurs : l'enjeu de la crédibilité. *Revue française du marketing*, 244/245, 45–60.

Delmas, M. and Burbano, V. (2011). The drivers of greenwashing. *California Management Review*, 54(1), 64–87.

Delmas, M. and Colgan, D. (2018). *The Green Bundle: Pairing the Market with the Planet*. Stanford University Press, Redwood City.

Du, S., Bhattacharya, C.B., Sen, S. (2010). Maximizing business returns to corporate social responsibility (CSR): The role of CSR communication. *International Journal of Management Reviews*, 12(1), 8–19.

Gabriel, P. (2003). Le marketing comme moyen de légitimation des entreprises dans une perspective de développement durable. *Décisions Marketing*, 29, 67–76.

Grant, J. (2009). *Le manifeste du marketing vert*. Afnor, Saint-Denis.

Jahnich, M. (2013). "S'il vous plaît… dessine-moi le marketing responsable !", Une étude exploratoire sur l'intégration du développement durable dans les stratégies marketing des entreprises. Report, ADEME [Online]. Available at: https://www.reussir-avec-un-marketing-responsable.org/sites/default/files/ademe_etude_marketing_responsable_entretien_katia_prassoloff-2012.pdf.

Laville, E. and Florentin, A. (2015). La marque positive : comment faire de la RSE un booster de vos marques ? [Online]. Available at: https://www.slideshare.net/Utopies_Paris/la-marque-positive-comment-faire-de-la-rse-un-booster-de-vos-marques.

Mauléon, F. and Silva, F. (2009). États des lieux de la RSE et du développement durable en France. *Management Avenir*, 3, 23–35.

Monnot, E. and Reniou, F. (2013). "Ras le bol d'entendre parler d'écologie !" : comprendre la contestation des discours écologiques par les consommateurs. *Décisions Marketing*, 71, 93–109.

Morsing, M., Schultz, M., Nielsen, K.U. (2008). The "Catch 22" of communicating CSR: Findings from a Danish study. *Journal of Marketing Communications*, 14(2), 97–111.

Newell, S.J. and Goldsmith, R.E. (2001). The development of scale to measure perceived corporate credibility. *Journal of Business Research*, 52, 235–247.

Ottman, J.A., Stafford, E.R., Hartman, C.L. (2006). Avoiding green marketing myopia: Ways to improve consumer appeal for environmentally preferable products. *Environment: Science and Policy for Sustainable Development*, 48(5), 22–36.

Robert-Demontrond, P. and Joyeau, A. (2010). Résistances à la responsabilité sociale des entreprises : de la critique des modalités à la critique de la logique économique sous-jacente. *Revue de l'organisation responsable*, 5(1), 5–16.

Schlegelmilch, B. and Pollach, I. (2005). The perils and opportunities of communicating corporate ethics. *Journal of Marketing Management*, 21(3/4), 267–290.

Vadarajan, P.R. and Menon, A. (1988). Cause-related marketing: A coalignment of marketing strategy and corporate philanthropy. *Journal of Marketing*, 52(3), 58–74.

van de Ven, B. (2008). An ethical framework for the marketing of corporate social responsibility. *Journal of Business Ethics*, 82(2), 339–352.

White paper (2016). Réussir avec un marketing responsable [Online]. Available at: https://reussir-avec-un-marketing-responsable.org/le-livre-blanc.

7

The Effectiveness of "Provocation" in Environmental Advertising: Beware of "Greenbashing"

Sihem DEKHILI[1] and Samer ELHAJJAR[2]

[1] CNRS – BETA, University of Strasbourg, France

[2] University of Balamand, Tripoli, Lebanon

7.1. Introduction

The emergence of a new generation of responsible consumers concerned about the environment is pushing companies to reconsider their business models and implement environmental initiatives. As a result, numerous brands in different sectors of activity are on the path towards a "green" status (Erdman 2008).

In order to valorize their responsible actions, companies choose environmental advertising. This is a tool aimed at promoting environmentally-friendly offers, informing and reassuring consumers of the ecological quality of products and encouraging responsible purchases. By means of such a communication strategy, companies are seeking, above all, to differentiate themselves from the competition and gain economic advantages (Banerjee *et al.* 1995). These methods of communication are more than necessary, given that ecological products are credence goods which are characterized by a high cost in the search for information on their

Marketing for Sustainable Development,

environmental quality. Consumers are never sure about the product's ecological characteristics, even after consuming it (Van Amstel *et al.* 2008).

However, opportunistic behaviors have emerged in this area: companies have attempted to spread distorted information about their actual environmental performance, leading to what we call "greenwashing" (Font *et al.* 2012). A recent example that attracted much media attention and scandalized consumers was related to Volkswagen. Between 2009 and 2015, the brand, which benefited from a favorable image, used fraudulent software to falsify cars' emissions levels during certification tests. According to a study carried out by TNS Global (2009), almost 90% of French people do not believe in the sincerity of companies' commitments and see them as merely being a commercial tactic. Consumers perceive environmental advertising as deceptive (Matthes and Wonneberger 2014), and discussions on forums (for example, *Marre de l'écologie à tout bout de champ*), as well as the increasing number of actions taken by various non-governmental organizations (NGOs) (for example, Friends of the Earth – France, known locally as *Les amis de la terre*), show a move towards questioning environmental advertising.

To overcome these limitations, companies have opted for a new type of communication, known as "greenbashing", which differs from traditional environmental advertising through the use of an ironic and provocative tone intended to criticize ecology (Elhajjar and Dekhili 2016). This mode of communication was started by Volkswagen[1] in 2009, during the launch of its environmentally-friendly Passat SW BlueMotion model (119 g of CO_2/km and 4.5 l of fuel/100 km), and was then adopted by other major brands, including Goodyear, Maxwell and Honda. The latter was noted for its famous slogan "became EGO-responsible", which was associated with the launch of its "first pleasure sport hybrid coupe".

However, greenbashing has offended some consumers; an advertisement that promotes a product's environmental benefit while mocking ecologists and green trends can be challenged! This raises the issue of the effectiveness of provocative environmental advertising. With literature on this topic being very limited, this chapter aims to explore the effect of provocation on consumer attitudes towards the promoted product and environmental advertising, on the one hand, and on the

1 An extract from an advertisement that uses an ironic tone and provocative images: "In founding ATMOS (an isolated village), our aim has been to live without rejecting a single gram of CO_2; we have renounced fossil fuels, refused all manufactured products, banned everything that comes from outside, because of transport, and have even stopped lighting fires. We have thus achieved our ideal!"

perceived credibility of the message and the ecological image of the advertiser, on the other. Our chapter will thereby offer a better understanding of greenbashing and its effectiveness among the target audience, and determine the relevence of such a communication strategy. It will be particularly important to examine the extent to which greenbashing could overcome consumer skepticism towards environmental advertising and mitigate the phenomenon of greenwashing.

7.2. Greenbashing: clarification of a new concept

7.2.1. *Advertising and contestation*

In marketing, the term "contestation" is not a new one; it reflects the act of disputing something and taking an unfavorable position towards it. Scholars (Marion 2003; Roux and Rémy 2009) have talked about the "contestation of marketing" in reference to criticisms addressing the discipline. The contestation in this case is focused on the usefulness and effectiveness of marketing, and on its use as a tool that tends to encourage consumerism. More specifically, some authors used the term "contestation of advertising" (Quesnel 1971; Piquet 2003; Granier 2008) to highlight the reluctance towards communications from brands. An anti-advertising movement has emerged since the late 1960s, considering advertising to be an invisible dictatorship over society and responsible for changes in individual behaviors (Ellul 1970). In this sense, advertisers, such as David Ogilvy, founder of the Ogilvy agency, have questioned the usefulness of advertising: "Does advertising increase prices? Does it promote monopolies? Can it make a consumer buy a lower-quality product? Does it make people buy products they don't need? Is advertising a web of lies?"

In summary, the main criticisms of advertising relate to:

1) its capitalist aspect: advertising reduces the individual to a simple consumer function and encourages consumption. It does not always provide all of the information on the manufacturing process and the characteristics of the offer, but primarily seeks to sell. The role of advertising is to suggest that individuals cannot live without buying, thereby increasing the power of consumption as a tool for social integration: "Our society thinks and talks like a consumer society. At least as much as it consumes, it consumes itself as a consumer society, as an idea. Advertising is the triumphal ode to this idea" (Baudrillard 1979, p. 312);

2) its moral aspect: the most common criticism of advertising is a moral one. This concerns, in particular, its potential impact on the behaviors of minors and the risk of manipulating the consciousness and promoting discriminatory stereotypes

(Gauthier 1997). A number of advertisements have been denounced for their contribution to child obesity. The polemic began in England, where the excessive presence of famous footballers in advertisements promoting confectionary and fast foods, aimed at young audiences, was deemed irresponsible. Another example concerns advertisements that encourage sexism. *La Meute*, an international mixed feminist network that has been engaged in combating sexist advertising since 2000, organizes demonstrations in places that disseminate advertisements that are considered to be problematic;

3) its environmental aspect: advertising has a considerable environmental impact. It is associated with visual and noise pollution and with significant energy expenditure. More specifically, advertising invades public spaces, such as public transport, and uses significant amounts of paper. It is also often criticized for its use of polluting ingredients, such as ink.

For these reasons, some authors (Cooper and Kelleher 2001) consider it a real challenge for advertisers to be ethical. In the case of ecological offers, beyond the characteristics of the products, which have to take environmental and social considerations into account, communications should also be held responsible.

7.2.2. *Environmental advertising: from greenwashing to provocation*

The concept of environmental advertising was introduced in the early 1990s to highlight advertising messages using environmental arguments to sell biodegradable, recycled or recyclable products (Kangun *et al.* 1991). Environmental advertising thus offers the receiver information on the environmental performance of the company and the ecological characteristics of the product (Polonsly *et al.* 1997). By explicitly or implicitly addressing the relationship between the goods (product or service) and the biophysical environment, it promotes an ecological lifestyle and contributes to "greening" the image of a company engaged in sustainability (Banerjee *et al.* 1995). However, environmental advertising has failed in conveying the real environmental benefits of the brand (Chan 2004) and in providing an honest ecological image (Peattie and Crane 2005). Various criticisms have noted, in particular, the ambiguity of the content of the messages, synonymous with a lack of effectiveness of such advertisements (Manrai *et al.* 1997).

Greenwashing practices have grown exponentially in recent years; some companies have been criticized for having spent more money on advertising than on the cause itself (Hoeffler and Keller 2002). Greenwashing is intended not only to sell more goods by exaggerating their ecological quality, but also to conceal negative information about the environmental image of the company and, consequently, to erect a kind of "green" smokescreen in front of consumers' eyes (Furlow 2010). Various studies have underlined the negative impact of greenwashing on consumer reactions: negative perceptions of misleading messages (Chen and Chang 2013), and a contestation of the products, brands and environmental claims conveyed in the advertising (Furlow 2010). The contestation of advertising can take different forms that vary in accordance with the degree of contestation. First of all, individuals display doubts about the senders of the environmental message and their intentions, which often leads to a loss of trust in the brand and its commitments. Beyond this, individuals may criticize and condemn any information conveyed that they consider manipulative. When the opposition is strong, the contestation of environmental advertising may take the form of resistance. In such cases, the message that is disseminated is considered a form of pressure, and the contestation is strong (Monnot and Renniou 2013; Elhajjar and Dekhili 2016).

When actors with varying degrees of sincerity convey the same message (here an environmental advertisement), the receivers are unable to distinguish which advertisers are serious and which are simply opportunistic. This situation leads consumers to question all companies. In such cases the message is no longer a guarantee for the target audience (Kirmani and Rao 2000). More naïve consumers, however, risk being deceived by a company that is acting in bad faith by attributing environmental qualities to it that it does not have (Boyer 2006).

In order to overcome the limitations of greenwashing and consumer skepticism towards ecological claims, some companies, bucking the general trend, have chosen to adopt a communication style that is based on criticism of the green movement and tinged with sarcasm and provocation (Shapiro 2009). Provocation is defined as a deliberate appeal, within the content of an advertisement, to stimuli intended to shock at least part of the audience, owing to associations with values or taboos that go unchallenged in the advertisement and/or because they are of a distinctive or ambiguous nature (Vézina and Paul 1997). There are thus three key characteristics linked to provocation: difference or originality, ambiguity in the advertising message and the transgression of norms by using taboos.

Various works in marketing literature (Vezina and Paul 1997; Laadhar and Romdhane 2013) have shown that provocative advertisements have a negative impact on consumer attitudes towards the advertisement and the product promoted, and on the perceived credibility of the advertisement, which often harms the brand image and thus leads to rejection of the advertisement. Moreover, Latour *et al.* (1990) highlighted the emotional responses generated by provocative advertising that influence individual attitudes towards the advertising in question.

7.2.3. *Greenbashing: what are the specificities of environmental advertising?*

While traditional environmental advertising is characterized by argument-based content that highlights a company's environmental engagement and the ecological benefits of its offer, greenbashing, on the other hand, is distinguished by its sarcastic content. The provocation in this case consists of deliberately mocking activists and/or green trends: "These advertisements present environmental activists trying to promote an ecological product in a ridiculous manner by mocking their speeches and their expectations for a world free of ecological problems"[2] (Solange Hemerey, consultant). In order to do this, greenbashing advertisements often play with words: "In French, Honda, for example, used the word "*écoresponsable*" ["environmentally responsible"] to create the term "*ego-responsable*" ["responsible for yourself"] for use in a slogan" (Lisa Buono, communications manager, Novamex L'Arbre Vert).

Companies can be pushed to adopt a greenbashing approach for various reasons. Indeed, this type of advertising can be of interest to companies in a sector with strong competition. Faced with competitors that are engaged in pursuing innovations in sustainable development, some companies, lacking the ability to catch up with the leaders in the field, choose to ridicule environmental issues in order to insinuate that ecological concerns are of little importance. Other companies use greenbashing to avoid the risk of being accused of greenwashing.

Provocative environmental communication seems to have several advantages. These include its ability to capture the attention of the target audience:

> Consumers don't like advertisements anymore and companies know it: advertisers are trying to attract consumers and, in my opinion, these advertisements have been able to do that. We can see the numbers of views that greenbashing advertisements have on YouTube, more than 200,000, while other conventional advertisements only have 1,000 views. (Alexandre Pasche, consultant)

The sarcastic tone particularly helps to make the advertising memorable: "Companies expect to launch a memorable advertisement because being memorable is one of the key criteria for the effectiveness of advertising" (Daniel Luciani,

---

2 The quotations in this section come from a qualitative study carried out on a sample of 23 advertising professionals. The sample was formed of communications managers, communications consultants and the heads of advertising agencies.

director of the Icom agency). Bored by traditional environmental advertising, consumers are finding a fresh approach in greenbashing advertising:

> Consumers have started to express a kind of weariness towards ecology; we can see it on social media, and even some politicians have expressed this attitude. I think that greenbashing is a company response to this weariness and an attempt to exploit it. (Charlotte Arnal, director of the PatteBlance agency)

Another advantage that can be associated with greenbashing is the ability for companies to differentiate themselves and create a buzz around their brands: "Companies want to launch an advertising campaign that could provoke society and make them famous" (Daniel Luciani, director of the Icom agency).

Box 7.1, based on the point of view of Christophe Bultel, an advertising professional, completes this section on the specificities of greenbashing.

According to Christophe Bultel, traditional environmental advertisements are often challenged. Negative reactions from the target audience can harm the brand in such cases: "The contestation damages the reputation and equity of the brand, and reduces consumer trust: less consumer trust means less loyalty to the company."

As a result, a number of companies around the world have opted for greenbashing campaigns, characterized above all by a provocative tone: "Sarcasm towards environmentalists… greenbashing thrives on overexposure to sustainable development and on the weariness of some consumers towards it." In their design, such advertising is primarily focused on mockery of the green movement and presents a description of the ecological characteristics of the product being promoted: "We always see a description of the product in the last part of this type of advertisements and, more particularly, the environmental features of the product and its benefits for the environment."

For some managers/marketers, this communication style is a solution that allows them to escape greenwashing: "Consumers currently tend to perceive all companies that talk about the environment as greenwashers. Greenbashing advertisements may be used because companies are scared of being accused of greenwashing. It is a way of presenting companies' environmental practices through a new type of advertising that is very different to traditional environmental advertising."

However, mocking environmentalists can be deemed as irresponsible as it can then be considered to be against sustainable development, which can have a negative impact on the image of the brand and the sincerity of the company: "They mock environmentalists and it isn't very responsible to criticize a large community… [This type of communication] doesn't promote responsible behavior, and there is a real risk of sending a message to citizens that they should slow down their efforts in resolving environmental problems."

Lastly, Christophe Bultel gives an opinion that, to him, sums up efficient communication: "I think that consumers need a sincere, educational type of communication that promotes the product and, of course, accurate messages that do not exaggerate the environmental performance of the product."

Box 7.1. *Interview with Christophe Bultel, Director of the RC2C PR agency*

7.3. The effects of provocation on the effectiveness of environmental advertising

7.3.1. *The empirical study: an experiment with consumers*

Our objective is to provide a better understanding of the effectiveness of provocative environmental advertising. More specifically, we propose to determine the effect of provocation on: 1) consumer attitudes towards the advertising, 2) consumer attitudes towards the product promoted, 3) the perceived credibility of the advertising and 4) the perception of the ecological image of the advertiser. The variables used come from the general model of attitudes to advertising developed by MacKenzie and Lutz (1989).

Regarding provocation, four possibilities have been included: 1) an environmental message that mocks ecological activists, 2) a message that mocks ecological trends, 3) a message using word play and 4) a non-provocative message as a control for the experiment.

According to the literature (Banerjee *et al.* 1995; Delmas and Burbano 2011), the source of the advertising can have an impact on the evaluation of it; individuals perceive environmental advertisements from NGOs more positively than those put out by companies. As a result, in this experiment we have taken the source of the environmental advertisement into consideration (NGO versus company).

By analyzing numerous environmental advertisements used on the market, we have created advertising posters for a fictitious brand of car tires that we have called "Pneuxo". We have also opted for a fictitious environmental NGO (the French Association for Ecology) (see Box 7.2).

In this study, we conducted a between-subject experiment. The independent variable is "provocation in environmental advertising", which has been used by means of four stimuli: mocking ecological activists, mocking ecological trends, word play and the absence of provocation in the message (control group). One moderating variable (the source of the advertising) has been manipulated by taking two modalities into consideration: company (Pneuxo) and NGO (French Association for Ecology). Moreover, control variables (age, gender, environmental concern) have been included.

The eight experimental conditions listed below (Table 7.1) have been considered (see Figure 7.1 for an example of an advertising poster proposed to respondents).

	Provocation in environmental advertising			
Source of advertising	Mocking activists	Mocking ecological trends	Word play	No provocation
Company	Group 1	Group 2	Group 3	Group 4
Association	Group 5	Group 6	Group 7	Group 8

Table 7.1. *The experimental conditions considered in the study*

After a pretest involving 11 communications professionals with the overall aim of testing the provocation-related stimuli, 320 consumers were randomly assigned to the eight experimental conditions described above.

The study was carried out in Strasbourg and the sample respected the city's proportions regarding gender, age and socioeconomic category.

All of the dependent variables were measured thanks to established scales from the literature: attitude towards the advertising (MacKenzie *et al.* 1986, 3 items), attitude towards the product promoted (MacKenzie *et al.* 1986, 3 items), perceived credibility of the advertisement (MacKenzie *et al.* 1989, 5 items) and perception of the ecological image of the advertiser (Manrai *et al.* 1997, 6 items).

All of the measurements used are highly reliable, with a Cronbach's alpha higher than 0.7. In terms of analysis, tests for comparing means, correlations and variance analysis (ANOVA) were performed.

Box 7.2. *Methodological details for measuring the effect of provocation in environmental advertisements*

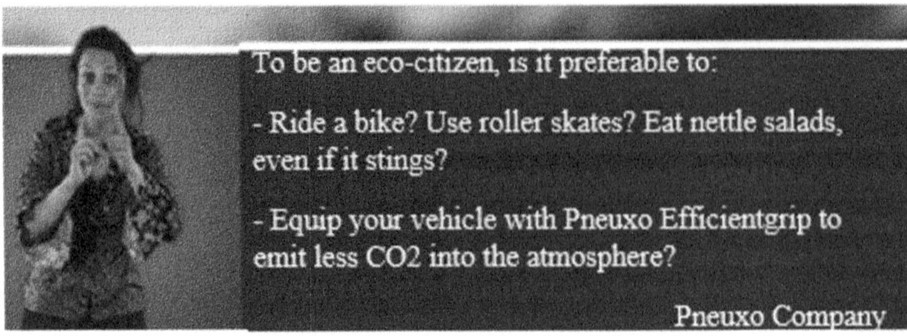

Figure 7.1. *Example of an advertising poster issued by a company mocking ecological trends*

7.3.2. *Effect of provocation on the effectiveness of environmental advertising: mixed results*

To test the effect of provocation on: 1) consumer attitudes towards the advertising, 2) consumer attitudes towards the product promoted, 3) the perceived credibility of the advertising and 4) the perception of the ecological image of the advertiser, comparisons have been made between the test groups exposed to provocative advertisements and the control group that saw a non-provocative advertisement.

Table 7.2 provides the mean values obtained from the test groups and the control group. It shows significant differences, except in the case of the attitudes towards the product promoted. Contrary to the conclusions of some previous works (Kirchler and De Rosa 1998; Laadhar and Romdhane 2013), it was found that provocation has no effect on attitudes towards the product promoted in the environmental advertising. The absence of a significant effect can be attributed to the nature of product tested (tires), which belongs to the category of utilitarian goods.

However, provocation has a negative effect on attitudes towards advertising, the perceived credibility of the advertisement and the perception of the ecological image of the advertiser. These results confirm the overall negative effect of provocation underlined by some authors (Manceau and Tissier-Desbordes 1999). As mentioned by Laadhar and Romdhane (2013), provocative advertisements seem to have a negative impact on the cognitive responses of consumers and, consequently, the perceived credibility of the advertising. Moreover, greenbashing advertisements do not provide consumers with tangible proof that allow them to objectively assess the environmental quality of the product and the concrete impact of actions put in place by the company, which may explain consumers' reluctance towards this style of communication. Consumers prefer environmental advertising that is precise, informative and detailed, helping them to make their purchasing decisions in relation to ecological products (Alniacik and Yilmaz 2012). Environmental claims should be able to encourage responsible lifestyles and improve the responsible image of the company (Banerjee *et al.* 1995; Deephouse and Carter 2005).

The dependent variables measured*	Advertising that mocks activists	Advertising that mocks ecological trends	Advertising that uses word play	Non-provocative advertising
Attitude towards advertising (7-point scale)	3.25 (1.87)	3.15 (1.75)	2.72 (1.62)	4.41 (2.12)
Attitude towards the product promoted (7-point scale)	3.85 (1.55)	3.92 (1.93)	3.75 (1.46)	3.68 (1.8)
Perceived credibility of the advertising (5-point scale)	2.38 (1.12)	2.29 (1.09)	2.1 (0.99)	3.24 (1.3)
Perception of the ecological image of the advertiser (5-point scale)	3.33 (0.99)	3.25 (0.99)	3.47 (1.03)	2.4 (1.15)

* All scales range from 1 (negative tendency) to 5 or 7 (positive tendency), except for the perception of the ecological image of the advertiser, where the scale is reversed.
The standard deviations are presented in brackets.

Table 7.2. *Mean values of dependent variables in the test groups and the control group*

We also aim to determine whether the source of the advertising (NGO versus company) has an effect on consumer reactions to greenbashing. Analyses have shown that the respondents express the same attitude towards the provocative advertising and the product promoted irrespective of the source. However, there is a significant difference in the perceived credibility of the advertising ($F\ [1.319] = 23.06$, $p<.001$): a provocative advertisement from an NGO was considered to be more credible than one from a company (M_{NGO} = 3.53 and $M_{Company}$ = 2.78). In addition, the perception of the ecological image of the advertiser depends on the source of the advertising message ($F [1.320] = 8.50$, $p<.001$): the use of provocation in an advertisement has a less unfavorable impact on the ecological image of an NGO than on that of a company (M_{NGO} = 2.63 and $M_{Company}$ = 2.84)[3]. Consumers are in fact increasingly skeptical of the environmental practices of brands (Do Paço and Reis 2012). The phenomenon of greenwashing is doubtlessly one of the reasons for this; it has harmed the image of companies, including those that have not been accused (Tremblay 2007).

Lastly, the analysis of the effect of the control variables indicated that the age, gender and environmental concern of consumers have no significant effect on the evaluation of provocative advertisements. While previous studies confirmed that there are no differences between men and women (Do Paço and Reis 2012), or between older and younger people (Boyer 2006) in their responses to advertising, there needs to be further examination of the result indicating the absence of an effect on environmental concerns. Some authors (Mars and Menivelle 2012; Richards 2013) asserted, in effect, that environmental concerns have an impact on consumer responses to environmental advertising. One possible explanation for this result is related to the composition of the sample used, as 3/4 of the people contained in the sample declared themselves to be concerned by environmental issues.

7.4. Conclusion

The chapter proposed to explore the effectiveness of provocative environmental advertisements among consumers. In greenbashing campaigns, brands saw a means of escaping suspicions of greenwashing, a way of differentiating themselves, or even a tool for capturing the attention of consumers who were weary of environmental discourse. However, the results obtained indicate that it is not a particularly valuable style of communication; provocation has a negative impact on attitudes to the advertising, the perceived credibility of the advertisement and the perception of the ecological image of the advertiser. The role of environmental advertising is not questioned in any way, but consumers simply manifest their desire to find more informative and truthful elements in the messages. While the provocative and sarcastic tone seems inconsistent with the defense of the environmental cause, other communications styles, such as humor, should be tested.

3 The scale for measuring the perception of the ecological image of the advertiser is reversed in comparison to the other scales considered, that is, the higher the mark, the less favorable the perception.

Concerning companies that have chosen greenbashing because they have failed to obtain a favorable position in terms of ecological engagement in comparison to their competitors, communication based on humility and honesty is, without doubt, a better option. Exaggerated and vague environmental claims (such as 100% green, 100% natural, no CO_2) that are used to capture the consumer's attention are often perceived very negatively by consumers and constitute a risk to the reputation of a brands (Delmas and Burbano 2011).

Brands have an interest in informing the target audience that they have not made the progress required in a given year, but that they plan to do better in the coming years. Humility and honesty also require the company to admit the mistakes they have made in past advertising campaigns. This is what happened with Renault, who declared that the mention of an "ecological character" in its advertising for the Zoé model, without putting it into context, countered the commitments it had made as part of the French Union of Advertisers by signing a responsible communication charter. As a result, the advertisement was changed by replacing the word "ecological" to "zero emissions". Moreover, the message becomes more credible when it is realistic, avoiding the overestimation of the ecological aspect of the product promoted.

Companies must avoid declaring themselves as the best on the market in terms of sustainability; stakeholders will still find fault with ecological practices. Moreover, brands should try to prove their maturity to their target audiences by owning up to the weaknesses in their engagement. They should avoid giving consumers the impression that it is possible to become "green" in the blink of an eye.

In order to increase the credibility of their claims, brands need to use tangible proof (Mikol 2003). Proofs consists of a set of elements made available to receivers that aim to defend the promise conveyed. For example, the brand Le Chat was severely criticized for the communication campaign it launched in 2009, which pur forward a baseless environmental claim: "It's time to talk less about the environment and do more about it". Five years later, the company came back with a revised message around the *Le Chat Eco-Efficacité* product: "the Ecolabel guarantees a wash that's tough on stains and soft on the planet", which highlights the European Ecolabel adopted by the brand.

Our results showed that provocation in environmental communications has a less unfavorable influence on the perceived credibility of the advertising and the perception of the ecological image of the advertiser when the advertisement comes from an NGO (as opposed to a company). More generally, NGOs benefit from perceived greater legitimacy and sincerity in relation to environmental actions. On a practical level, it will be fruitful for brands to create partnerships with associations who can provide their expertise as well as their lobbying powers (Ollitrault and Chartier 2006, pp. 93–115).

Lastly, the communication style should be analyzed within the framework of a global approach of responsibility. All aspects of the communication campaign should be considered as a whole. The example of the entirely eco-designed campaign of Biocoop is very interesting in this regard. As a retailer of organic products, it reaffirms its ecological commitments by presenting a film, five images and a website illustrating its values and environmentally responsible approach. Less polluting means of transport were used to get to the various filming locations. The traditional design and production processes were replaced by environmentally-friendly solutions. In addition, the website launched was designed to be as light as possible (it weighs only 3 MB). The photos were taken with a 40 by 50 cm pinhole camera made of old wooden crates, and were developed on-site using recycled products. In total, the ecological footprint of the Biocoop campaign was reduced to a third, i.e. 5.9 tons of CO_2, versus the 15.2 tons that would be generated in the case of an identical campaign produced in a conventional manner.

7.5. References

Alniacik, U. and Cengiz, Y. (2012). The effectiveness of green advertising: Influences of claim specificity, product's environmental relevance and consumers' pro-environmental orientation. *Amfiteatru Economic Journal*, 14(31), 207–222.

Banerjee, S., Gulas, C.S., Iyer, E. (1995). Shades of green: A multidimensional analysis of environmental advertising. *Journal of Advertising*, 24(2), 21–31.

Boyer, J. (2006). Le scepticisme du consommateur face à la publicité : modèle intégrateur et effets du signal satisfait ou remboursé. PhD thesis, Pierre Mendès-France University, Grenoble.

Chen, Y.S. and Chang, C.H. (2013). Greenwash and green trust: The mediation effects of green consumer confusion and green perceived risk. *Journal of Business Ethics*, 114(3), 489–500.

Deephouse, D.L. and Carter, S.M. (2005). An examination of differences between organizational legitimacy and organizational reputation. *Journal of Management Studies*, 42(2), 329–360.

Delmas, M.A. and Cuerel Burbano, V. (2011). The drivers of greenwashing. *California Management Review*, 54(1), 64–87.

Do Paço, A.M.F. and Reis, R. (2012). Actors affecting skepticism toward green advertising. *Journal of Advertising*, 41(4), 147–155.

Elhajjar, S. and Dekhili, S. (2016). Could the greenbashing be a solution for the environmental advertising failures? *Journal of Marketing Trends*, 3(1), 11–19.

Erdman, B. (2008). Is green really your color? *Brandweek*, 49(5), 18.

Font, X., Walmsley, A., Cogotti, S., McCombes, L., Häusler, N. (2012). Corporate social responsibility: The disclosure–performance gap. *Tourism Management*, 33(6), 1544–1553.

Furlow, N.E. (2010). Greenwashing in the new millennium. *The Journal of Applied Business and Economics*, 10(6), 22.

Hoeffler, S. and Keller, K.L. (2002). Building brand equity through corporate societal marketing. *Journal of Public Policy and Marketing*, 21(1), 78–89.

Kangun, N., Carlson, L., Grove, S.J. (1991). Environmental advertising claims: A preliminary investigation. *Journal of Public Policy and Marketing*, 47–58.

Laadhar, F. and Romdhane, R. (2013). Provocative persuasive advertising and mediating role of emotions: A comparative study of two provocative ads. *Interdisciplinary Journal of Contemporary Research in Business*, 5(2), 622–640.

LaTour, M.S., Pitts, R.E., Snook-Luther, D.C. (1990). Female nudity, arousal, and ad response: An experimental investigation. *Journal of Advertising*, 19(4), 51–62.

MacKenzie, S.B. and Lutz, R.J. (1989). An empirical examination of the structural antecedents of attitude toward the ad in an advertising pretesting context. *Journal of Marketing*, 53(2), 48–65.

Manrai, L.A., Manrai, A.K., Lascu, D.N., Ryans, J.K. (1997). How green-claim strength and country disposition affect product evaluation and company image. *Psychology and Marketing*, 14(5), 511–537.

Mars, M.C. and Menvielle, L. (2011). L'influence du contenu du message publicitaire et des caractéristiques motivationnelles sur les réponses du consommateur aux publicités vertes. *10th International Marketing Trends Conference*, Paris, France, 20–22 January.

Mikol, A. (2003). La communication environnementale de l'entreprise. *Revue française de gestion*, 6, 151–159.

Monnot, E. and Reniou, F. (2013). "Ras le bol d'entendre parler d'écologie !" : comprendre la contestation des discours écologiques par les consommateurs. *Décisions Marketing*, 71, 93–109.

Ollitrault, S. and Chartier, D. (2006). ONG et développement durable : les liaisons dangereuses. In *Le développement durable : enjeux politiques, économiques et sociaux*, Aubertin, C., Vivien, F.-D. (eds). La Documentation française, Paris.

Peattie, K. and Crane, A. (2005). Green marketing: Legend, myth, farce or prophesy? *Qualitative Market Research: An International Journal*, 8(4), 357–370.

Richards, L. (2013). Examining green advertising and its impact on consumer skepticism and purchasing patterns. *Elon Journal of Undergraduate Research in Communications*, 4(2), 78–90.

Vezina, R. and Paul, O. (1997). Provocation in advertising: A conceptualization and an empirical assessment. *International Journal of Research in Marketing*, 14(2), 177–192.

8

How Can We Communicate Effectively About Climate Change?

Philippe ODOU, Marie SCHILL and Manu NAVARRO
REGARDS, University of Reims Champagne-Ardenne, France

8.1. Introduction

It is not easy to talk about climate change. Words are important and indicative of the way in which climate issues are seen. Should we talk about global warming, climate change or climate disturbance? Talking about global warming is certainly reductive as the climatic consequences of the resulting changes are varied, and include warming air, droughts and floods. Other consequences that are much less known among the general public have recently been observed, such as an increase in the acidity of the oceans. Moreover, if consequences relating to the displacement of populations or lower agricultural yields are taken into account, we can also talk about a global crisis of environmental, economic, social and political dimensions.

Beyond concerns about which name reflects the scale of the problem, as well as changes to the way it is perceived by the public, the climate crisis is largely dependent on the mobilization of public authorities. Citizens are involved in the sense that they support potentially restricting public policies, as well as consumers to the extent that they change their consumption habits. These elements are leading to

Marketing for Sustainable Development,
coordinated by Sihem DEKHILI.

changes in consumerist lifestyles, but also, and particularly, the transformation of the concept of the relationship between humankind and nature.

In this context, how can citizens and consumers be encouraged to engage further with the fight against climate change? Scientists and NGOs have tried to alert individuals with facts and alarming images. This strategy has proved to be effective in raising awareness, but broadly insufficient in its ability to change lifestyles and consumption habits. While the notable differences between countries, including within Europe, must be noted, the production of greenhouse gases has continually increased, owing to the economic development of China and India, to name just two of the most important examples.

Several theories, which we will go on to discuss in detail later, allow us to better understand why, despite an increasing threat, consumers and citizens are slow to change their behaviors. We will rely largely on Witte's Extended Parallel Process Model (EPPM) (1992). First, we note that individuals do not tend to believe that they have the individual or collective capacity to impact. The EPPM supposes that the perception of a highter threat tends to create defense mechanisms to control the fear rather than the actual threat; that is, these defense mechanisms control our emotions (control fear) instead of changing our behaviors (control the threat). Most studies show that individuals are aware of the problem and that it has only worsened in recent years, even in countries considered more resistant to change, such as the United States (see the studies and research of the CCAM[1] research project). According to an Ifop study for WWF[2], for all of the countries studied (France, the United States, Australia, Italy, Germany), with the exception of China, the risks linked to climate change are foremost among environmental concerns. However, most experts and researchers agree that this awareness has not been translated into significant behavioral changes to the extent needed (O'Neill and Nicholson-Cole 2009). Renouncing the comfort and well-being that comes with a materialistic lifestyle is not easy in hyper-consumerist societies.

Section 8.2 provides an assessment of the awareness of climate change and details the psychological mechanisms that could present obstacles to behavioral changes. Section 8.3 examines which communication should be favored and considers the discussion on the use of positive and/or negative emotions in communication on the topic of climate change. Lastly, section 8.4 concerns a key target group in communications strategies: children.

---

1 Climate Change in the American Mind.

2 This 2017 study is available on the WWF website.

8.2. A gap between awareness and behavior

8.2.1. *Awareness of the threat posed by climate change*

The majority of the studies carried out in recent years conclude that there is an increased awareness of the risks linked to climate change. This awareness translates into a greater level of concern for nuclear and food-related risks. Extreme weather events are the most feared, followed by issues of flooding and drought.

While climate skepticism remains, the reality of climate change is broadly accepted (around 63% of people in the United States and 79% in China believe in climate change) with a split between, on the one hand, the United States and Australia where doubt persists and, on the other hand, other countries where skepticism is fading. Moreover, the majority of citizens are already feeling the effects of climate change (74% of French people) and feel worried, even as differences remain as to whether it has been caused by humans (91% of people in China believe this and only 61% do in the United States, despite a strong increase in this figure since 2010, +11%)[3].

As the Ifop study shows, and the studies led by the CCAM research project in the United States confirm, the perception of the threat is changing, awareness is widespread and is making positive progress, following a period of stagnation and despite significant lobbying from some industries[4]. Recent demonstrations and "climate marches" confirm this impression.

It therefore seems that while awareness is a reality, it has not been sufficiently quick in leading to transformations in our behavior. This lack of effectiveness has been highlighted by numerous researchers, essayists and politicians for a number of years in their calls for a Green New Deal (Klein 2019). They have drawn attention to cultural and ideological resistance to communicating about climate change. In addition, it is not enough to simply convey scientific information about the reality of climate disturbances to convince individuals to change their beliefs, or even alter their behavior. Such resistance must be related to individual political convictions. In the United States, Conservatives tend to be more skeptical of climate change than Democrats and are therefore more likely to justify the existing system when it is challenged (Feinberg and Willer 2011).

---

3 Ifop study for WWF (2017).

4 In the United States, according to The New York Times, the industries most involved in the emission of greenhouse gases spent more than 1.35 billion dollars to lobby members of Congress between 2000 and 2016.

While awareness exists, it is not necessarily being converted into actions to combat climate disturbances. Most people are under the impression that they do not have to alter their behavior. Significant changes to modes of consumption often take a long time to establish. For example, people do not become vegetarian overnight and the time it takes to replace a fleet of vehicles or a household heating system is also relatively long. This being the case, there are also numerous obstacles to change that have been identified. We will go on to discuss the psychological obstacles, while remaining conscious that other social, cultural and political obstacles can limit change too.

8.2.2. *Psychological obstacles to changing our modes of consumption*

While our fellow citizens are, for the most part, aware of the dangers associated with climate change, there are numerous obstacles that can prevent this awareness being converted into direct action (mode of consumption) or indirect action (encouraging societal change by participating in political debates, demonstrations, by lobbying or by supporting limited political programs). We will detail the psychological obstacles by relying largely on Witte's EPPM, which identifies two key elements of the process of change in relation to a threat: perceived efficacy and perception of the threat.

Perceived efficacy: in most models related to behavioral change, the role of self-efficacy, which can be defined as an individual's trust that they have the ability to control their own behavior, is considered decisive. Beyond this control, there are other factors, such as the perceived effectiveness of the result (personal outcome efficacy) that an individual has in relation to the objective to be achieved, as well as collective outcome efficacy. Several issues can arise on the issue of efficacy alone.

In terms of perceived control, consumers may have the impression that they cannot do more because of a lack of means at their disposal (absence of public transport, etc.), a lack of financial means (inaccessibility of electric vehicles, etc.), a lack of time or a lack of autonomy in decision-making (communal living, etc.). Regarding outcome efficacy, people may tell themselves that their individual actions have little impact on the global problems facing our planet. Lastly, in regard to collective efficacy, there can be significant differences between countries with a strong collective culture, such as northern countries, and southern European countries. In northern countries, people tend to believe in the benevolent nature of their fellow citizens and people tell themselves that if they are personally acting in support of the environment, then their fellow citizens will too, while in southern European countries, engagement in combating climate change can be negatively affected by the idea that our peers are not taking action. We can therefore conclude that the efficacy of the result is partially linked to the idea that we have of collective efficacy, and that our perception of collective efficacy at the national and international levels also acts as a reason for a lack of engagement with the fight against climate change.

In short, an individual, even one convinced of the seriousness of the threat, cannot take action, either because they think that they do not have the means or ability (low personal efficacy), or because they do have the means but think that their personal action will have little impact (low outcome efficacy), or because, even in the unlikely event that everyone is actively combating climate disturbance, it is not enough to overcome a threat that seems inescapable (low collective efficacy).

This problem of perceived efficacy becomes ever more serious as the perception of the threat grows. As has been shown by Witte and Allen (2000), when the perceived threat is higher and individuals think that there are no effective solutions, they tend to respond with defense mechanisms that allow them to control their fear rather than the threat itself.

Perception of the threat: climate change is linked to the perception of a threat. Witte (1992) explains with the EPPM that when we perceive a threatening situation, we make two evaluations: (1) an evaluation of the threat in terms of perceived seriousness and weakness; and (2) an evaluation of the "perceived efficacy", that is, an evaluation of the efficacy of the proposed solution ("response efficacy") and an evaluation of our ability to handle this action well (feeling of "self-efficacy").

Three situations can thus be envisaged:

– threat perception that is too low to provoke any response in people;

– a threat linked to a feeling of being unable to cope, which will trigger a fear-control process (detrimental to the target behavior). This process can be responsible for the appearance of coping strategies such as denial, reactance and defensive avoidance[5] (Witte and Allen 2000);

– a threat linked to a strong feeling of being able to cope, which will trigger a danger-control process (beneficial to the target behavior). This third situation is desirable as it pushes individuals to reconsider their behavior so that it aligns with their motivation to protect themselves.

5 To control their fear rather than the threat, people can use different methods such as denial, which consists of thinking that the threat will not reach the individual, reactance, which consists of resisting attempts at persuasion, and defensive avoidance, which consists of making an effort to avoid thinking about it.

Identifying the major psychological obstacles in relation to the adoption of new behaviors must be considered a prerequisite for launching a communications strategy. Following on from these obstacles, section 8.3 summarizes the main contributions to marketing research in terms of effective communication about the fight against climate change.

8.3. How can we communicate about climate change?

8.3.1. *What kind of communication should be encouraged?*

Beyond the obstacles identified and that must be taken into consideration in communications strategies, by working simultaneously on threat perception and response efficacy, research also concerns the kind of communication used. Fairly naturally, communications on climate change feature criticism of materialist and consumerist cultures and their short and medium-term consequences, especially for the sustainable survival of humankind on our planet. However, an effective communications strategy requires discussions to be adapted for the target audience, and arguments for different target audiences must also be considered.

Maibach *et al.* (2008, p. 497) have thus called for the choice of the kind of communication suited to each section of the American population, by relying particularly on the cultural values of the target audience in question. In order to make the communication more acceptable for a conservative target audience, these authors suggest the use of different means on the basis of an economic argument ("this is an excellent way of saving money"), an argument of energy independence ("this is a way for our country to free itself from dependence on foreign oil"), a heritage-based argument ("this is a way of protecting our children's future"), a moral argument ("this is how I honor my moral obligation to protect the wonders and mystery of life"), a religious argument ("this is a way to serve God by protecting His creation"), a nationalist argument ("innovative technology will sustain the economic vigor of our country") or even a health-based argument ("this is a way for me and my loved ones to stay healthy"). These different kind of communication share the idea of using a positive framework for communicating about climate change, while the initial communications strategies tended to focus on losses linked to inaction. However, the emotions generated are positive, contributing to the creation of hope overall, but, on the other hand, they are rather negative and maintain environmental anxiety and fear. As we will see in the section on emotions and their relationship with communication regarding climate change (section 8.3.2), there is no one way of communicating about the subject, and these two approaches, far from being opposed, may be complementary depending on the level of environmental engagement and the political stance of the citizen/consumer.

In the face of ever more sophisticated communications strategies, a conservative counter-movement is developing with the aim of defending the system in its present form, and providing citizens with counter-arguments to polarize the political debate and provide reasons to resist attempts at persuasion. McCright and Dunlap (2010) demonstrate how the defenders of the current economic system have been mobilizing in recent years by relying on Republican media (Fox News), social networks and by lobbying, to create doubt around scientific knowledge and to propose an alternative view on the origins of climate change. This strategy proved to be temporarily effective in most English-speaking countries, such as Australia, the United Kingdom and, of course, the United States. However, according the most recent CCAM studies, it now appears to have reached its limits.

8.3.2. *Which emotions should be focused on in the fight against climate change?*

Whether in films (*Tidal Wave*, 2009, *The Day After Tomorrow*, 2004, etc.) or in documentaries (*An Inconvenient Truth*, 2006), the register used to communicate about climate change is broadly an anxiety-inducing one (natural disasters, flooding, fires, droughts, threat of the end of the world, the extinction of a species, the end of civilization). As highlighted by Akil *et al.* (2018), an explicit or implicit threat of death is often linked with communications about climate change, triggering a state of existential anxiety related to the disappearance of the self. That being said, in a "blackwashing" logic, that is, the use of dramatizations that some consider excessive, this register is often used by activists and NGOs aiming to grab the attention of citizens and alert them to the consequences of climate change. However, many researchers have issued warnings about the defensive reactions that can be triggered and which are detrimental to the cause (O'Neill and Nicholson-Cole 2009). According to the terror management theory (TMT), there is a tension between our awareness of death and the desire to live that is alleviated by defense mechanisms, such as denial, minimization of the threat and distraction (proximal defenses), or by enhancing the cultural vision of the world that the individual identifies with to increase their self-esteem (distal defense). In a materialist culture, it is therefore through the accumulation of goods and hyper-consumption that consumers reduce this anxiety. As a result, when communications bring death to the foreground, the consumer who identified with a materialist cultural vision will be encouraged to increase their level of consumption. On the other hand, Akil *et al.* (2018) show that the effect is reversed when these individuals have an environment-based cultural vision. In terms of recommendations, NGOs and actors active in the fight against climate change must be very careful in their use of images, such as those depicting natural disasters which put death in the foreground. In a predominantly materialistic culture, these are counterproductive except for individuals with a more environmentally-minded cultural view of the world.

As we have just discussed, highlighting the catastrophic consequences of climate change can have paradoxical effects, but what about messages that create fear without necessarily making explicit or implicit references to death? Broadly speaking, as we saw in section 8.3, research has been unanimous on this subject. A message based on a fear that provides no solutions and that allows the individual to control the threat, invariably makes the individual control their emotions by denying the fact that there is something to fear, or they reject the message. This strategy has been considered partially responsible for the development of resistance to environmental discussions and consequently for the polarization of debates on climate change. The difficulty of communication thus lies in warnings about the consequences of climate change to capture people's attention and ensuring that this message is not only taken seriously, but that it creates the kind of anxiety needed to change our beliefs and our behavior without causing denial or rejection.

In contrast to this approach of triggering negative emotions, many researchers are interested in the influence of hope (Ojala 2012). Such work shows that messages based on hope are more effective than messages based on fear and that they also lead to feelings of self-efficacy. More generally, in recent years they have helped to restore positive emotions that broadly, when we use the EPPM, make it possible to develop the perception that an effective response is possible. These two strategies for using emotions are probably less opposed than they first appear at first glance. After an initial phase where the challenge was to ensure that a threat that could seem distant and abstract was taken seriously, it is perhaps time to provide citizens with examples of concrete solutions that can be applied to their daily lives and that could collectively make a difference. Rather than put the two strategies in opposition, a more nuanced approach should be adopted, taking into consideration the fact that each of these strategies can be of interest depending on the level of awareness, interest and resistance of the target audience. Numerous variables can diminish or strengthen the influence of a communication on climate change, such as cultural views, as we already mentioned, political stance (which may be partially superfluous depending on the cultural view), gender, religious belief and environmental engagement, to name just a few of the main variables.

However, many correlational studies have been able to provide evidence of a strong relationship between positive and negative emotions and the pro-environmental attitudes of individuals (Schill and Odou 2017). The studies agree that the intensity of the emotions felt is strongly linked to the level of consumer engagement with the fight against climate change. The emotions most commonly seen, other than fear and hope, are moral emotions such as shame and guilt, but also anger (against states or individuals who are not taking action) and even pride in taking action in support of the environment. At this point, it is necessary to differentiate between emotions linked to climate change itself and emotions that people can feel or contemplate feeling in relation to behaviors or the absence of behaviors in the fight against climate change.

Regarding the regulation of behaviors, Baumeister *et al.* (2007) show that it is not the emotional responses experienced at any given time that have the greatest influence on future behavior, but rather those that the individual thinks they will have (anticipated emotional reactions) when they adopt the behavior in question. The emotional responses experienced, whether they are negative like regret, guilt or even shame, or positive, such as joy or pride associated with a specific behavior, are used as information to regulate future behavior. The emotions are memorized and, over time, they facilitate learning about what a behavior can trigger at the emotional level. The anticipation of emotional reactions influences decisions and thus makes it possible to guide behaviors by promoting those linked to positive emotional reactions ahead of those linked to negative emotional reactions.

In their research on the influence of emotions linked to climate change, Schill and Odou (2017) were able to identify the main positive emotional reactions linked to behavior that combats climate change, and the negative ones linked to the absence of action. Consumers mainly feel satisfaction and pride in taking daily action to combat climate change, such as Olivier[6]:

> Personally, it's a source of motivation to consume better, with a smaller footprint, and I take a kind of personal satisfaction from it, a kind of pride, and I find consistency between what I think and what I do.

When action is not taken on a daily basis, the emotional response most often noted is guilt in regard to future generations. Didier said:

> When we could feel guilty is when we consider our grandchildren, if we aren't doing what we should be doing.

6 In this research, we conducted 10 semi-directed interviews, including Olivier, 35, a bank manager, and Didier, 60, a doctor. The aim of this study was to identify the main emotional reactions linked to climate change. This qualitative study was then extended by a quantitative survey with a sample of 525 individuals.

Not everyone feels guilty, or the guilt is lessened by their belief that individual responsibility for such an enormous global phenomenon is low. Other emotional reactions linked to the actions of others have also been identified, such as anger or sadness towards individuals, companies or states that are insufficiently involved in the fight against climate change. The results of this research show that, in contrast to what is generally believed, the positive emotional reactions experienced in relation to action combating climate change, such as pride and satisfaction, are as influential as negative emotions, such as guilt and fear (Schill and Odou 2017). In other words, even if the emotions linked to climate change are negative overall, it is still the search for anticipated positive reactions rather than the avoidance of anticipated negative reactions that will drive actions in this struggle. The influence of anticipated emotional reactions varies in accordance with the six emotional profiles identified in this research (Table 8.1).

Emotional profiles	Sociodemographic characteristics	Intentions to act	Variables influencing intentions to act
Engaged	Women > 40	Strong	– Anticipated positive emotional reactions; – attitudes to combating climate change.
Resistant	Men < 25	Very little	– Anticipated negative emotional reactions; – perceived behavioral control.
Unmotivated	Men > 50	Little to very little	– Perceived behavioral control.
Indifferent	Women < 25	Little	– Perceived behavioral control.
Overlooked	Men 25 to 55	Little	– Anticipated negative emotional reactions; – attitudes to combating climate change; – subjective norms.
Positive	Women < 25	Significant	– Anticipated positive emotional reactions; – perceived behavioral control; – subjective norms.

Table 8.1. *Emotional profiles in relation to combating climate change*

The work by Schill and Odou (2017) leads us to propose two recommendations:

Strengthen anticipated positive emotional reactions: communications encouraging action against climate change initially sought to scare, as well as blame, individuals. This strategy is not of great interest, in part because the target audience is the most resistant to combating climate change. It seems of greater use to work on strengthening the anticipated positive emotional reactions in order to encourage the "positive" people (who form the majority) to take further action, and to encourage those who are "engaged" to continue taking action.

Do not forget to work on self-efficacy (or perceived behavioral control): this variable has the greatest influence on intentions to combat climate change. It is also very important to different target audiences, whether they lack motivation or are positively inclined. It is therefore necessary to convince them that it is possible to take action by providing examples of concrete actions that are within an individual's reach.

While the vast majority of work on climate change concerns the perceptions, emotions, intentions and behaviors of adult populations, there is little research focusing on children. In section 8.4, we will examine issues related to the communication of climate change and children.

8.4. Mental representations of climate change among children

The vast majority of work on climate change is focused on adult individuals. There are many works that brandish children as future generations and the need to act in their interests. Young people – children and teenagers – are certainly the best to influence as the decisions they make regarding climate change will affect their future living situations. As prescriptors, consumers and citizens, young people are the central actors in the issue of climate change (Walker 2017). However, few works have examined this issue through the prism of children and adolescents, the latter being dependent on the decisions made by older generations. From this perspective, the younger generation is often considered passive and a victim of the negative effects of climate change (Haynes and Tanner 2015).

We have recently seen a change in this young generation. It is increasingly mobilizing, as shown by the manner in which schoolgirl Greta Thunberg has become a spokesperson, the global climate strike by young people on Friday September 20, 2019, and even the calls for school climate strikes. Faced with the climate emergency and its inevitable impacts on the young generation, the issue of communicating with that generation raises its head. In order to communicate best with the young generation, there must first be an assessment of the state of its engagement and of images of climate change, before proposing areas for reflection on the content and kind of communication. We will reflect on these areas of interest on the basis of work carried out among children aged 8 to 11.

8.4.1. *Engagement and representations of children relating to climate change*

Schill and Ezan (2018) carried out qualitative research among 42 French children in primary school (from CE2 to CM2 – the equivalent of year 4 to year 6 in the British system and third to fifth grade in the American system). The children all seemed to sing from the same hymn sheet: "climate change is the Earth getting warmer"; climate change "is when the temperature of the Earth gets higher". In other words, they all expressed that climate change corresponds to an increase in the temperature of the Earth's surface. The children also identified causes and consequences of climate change. Among the causes identified, the main one was pollution (cars, industrial activity, harmful human behavior such as smoking and littering). These causes thus seem relatively close to children from a spatial and temporal perspective. They acknowledged that these activities and behaviors create climate change and have harmful effects on the environment and humankind. The children thus identified illnesses and death, for both flora and fauna as well as for people. However, these mental images of the consequences of climate change, as opposed to the causes, seemed rather abstract and distant for the children. Mehdi (8) thought that the consequences affected children in other countries, while Oscar (8) and Maxime (9) thought about negative consequences for polar bears. Children therefore have mental images of climate change that vary between the concrete and the abstract. From a cognitive perspective, children seem engaged, regardless of whether their images of climate change were concrete or abstract.

At the emotional level, climate change creates emotional reactions that are expressed quite strongly among children. For the most part, these emotional reactions are negative, such as sadness. In this respect, the emotional reactions do not appear to differ greatly from those expressed by the adult population, with the difference that adult individuals express positive emotional reactions when they take action against climate change. Children drew an unhappy Earth, a crying sun or crying animals, or dead nature. They used relatively dark colors, such as gray and black, accentuating the feeling of sadness linked to this phenomenon. It thus seems that children, through the sadness they express, have a concrete emotional engagement with climate change.

From a behavioral perspective, few children drew behaviors to combat climate change. Only one child drew behaviors, such as separating trash, avoiding energy waste and even paying attention to the use of natural resources. Behavioral engagement thus seems relatively low among children and raises the question of their daily actions to combat this phenomenon, suggesting the adoption of a relatively distant and abstract view of the possible behaviors.

In summary, children seem cognitively and emotionally engaged in climate change, but little engaged from a behavioral point of view. On the basis of these findings and in keeping with the works of Dillahunt *et al.* (2017) and Ojala (2012) that highlight the need to reduce the psychological distance of children to encourage their engagement with the issue, we propose the following areas for consideration.

8.4.2. *How can we talk to children about climate change?*

Approach 1: develop and enhance the cognitive engagement of children: article 12 of the Paris Agreement (COP21) establishes the obligation to provide all pupils with education about climate change. In France, the Ministry for National Education is progressively integrating issues related to climate change, biodiversity and greenhouse gases into school programs at the primary and secondary levels. While this initiative is laudable, this integration is nonetheless being carried out in a fragmented and even disorganized manner. Providing the information to schoolchildren in a holistic and crosscutting manner would facilitate cognitive engagement. School programs with awareness-raising tailored to the variable cognitive stages of young people from kindergarten, and following a coherent program throughout the primary and secondary stages would enable children and teenagers to have a sufficient cognitive grounding to be able to understand the phenomenon. Outside of school, which seems to be the preferred agent of socialization for developing the cognitive engagement of the young generation, parent/child workshops could be established to transfer learning from schools to family settings.

Approach 2: support children in their emotional engagement: climate change can be an anxiety-inducing subject for children and the young generation more broadly, who are often emotionally defenseless (Ojala 2012). Children should thus be supported in order to suggest strategies to them that could enable them to confront such a serious and complex issue. For example, an excess of information from a very young age can risk creating negative emotional reactions among young children, which could last throughout their emotional development. Pessimism has also been shown to increase with age. As a result, this is a question of taking into account the emotional reactions of children and adolescents in the implementation of their cognitive learning. There are two major strategic positions that can be investigated in regard to the emotional dimension. The first consists of regulating the negative emotional reactions of children and teenagers. Detachment from the issue is relevant among the young generation. Young people seek to protect themselves from negative emotional reactions related to climate change by avoiding some information and talking about it to their social group of reference (friends, school, family). We note that amongst older young people, there is a strategy of over-activation of negative emotional reactions, emphasizing their distress and their powerlessness in the face of the scale of the issue. Reducing psychological distance in its emotional dimension by communicating information that creates negative emotional reactions can therefore prove to be difficult to manage for young people. The second strategy is a proactive one that considers more positive emotional reactions to future problems, such as hope. This second strategy relies largely on a greater psychological distance as it expresses hope for a better future. This strategy of putting climate change at a distance and communicating hope can prove to be constructive for young people. In summary, young people can be destabilized by the scale of climate change and can experience very negative reactions that they are ill-equipped to manage. Working on the emotions they experience and helping them to create distance from their emotional reactions could give them the tools they need to be able to take action when required.

Approach 3: reflect on ways of engaging children from a behavioral perspective: following on from reflections on the development of cognitive engagement and emotional support, it seems relevant to help young people become engaged in the fight against climate change from a behavioral perspective. Beyond knowing which are the "good" actions for combating the issue, this is a matter of making these actions part of the daily routine. In elementary school, these actions are most often reduced to learning how to sort trash. Going further, schools could put in place compost bins that the children could take care of on a daily basis. Schools could also develop initiatives to educate young people about responsible consumption by offering organic meals in the canteen or meals that come from sustainable agriculture. In terms of responsible consumption, schools could launch workshops to recover objects through DIY. At the high school level, technical education could be focused on repairing small materials. Scientific education could introduce young people to the manufacture of natural products to avoid the overconsumption of synthetic products (such as household products).

To compare the main research outcomes with the feelings of actors in the field, we called on Jacques-Olivier Barthes, Director of Communication at WWF France (Box 8.1). According to Barthes, the issue of climate change denial will one day be behind us and we are currently seeing a radicalization, especially among young people who have an acute awareness of the climate emergency. As far as children are concerned, Barthes recommends transparency in regards to the consequences of climate change. Positive communication on the current potential solutions is the preferred register of institutions. Barthes broadly supports the movement launched by Greta Thunberg to stigmatize the lifestyles and consumption habits of older generations. However, he admits that his view of things is greatly influenced by what he sees in Paris.

In your opinion, how has communication on climate change altered in recent years?

There's been a break since the arrival of Greta Thunberg. Since the Kyoto COP, it has been necessary to communicate in a positive manner so as to not cause people stress, as this creates a lack of reaction, or even apathy. This communication, as used by Al Gore, essentially said that all is not lost, to make people want to get involved, and then Greta Thunberg arrives and breaks the rules with rhetoric along the lines of 'we're not up to speed, we're going to be the generation that gets sacrificed, you should all be ashamed, this is a scandal'. The problem is that understanding doesn't necessarily lead to action, according to experts in cognitive science who have shown that our brain cannot respond to this type of threat: it's distant, it's not concrete, we're all responsible.

What's changing is that it is becoming concrete, even in Western Europe, which is leading to increased awareness among a fair few actors, and young people have a much more aggressive way of communicating about the responsibility of states and decision-makers. Between *The Times*' coverage of Al Gore in 2007 and of Greta Thunberg in 2019, there have been 12 years where nothing has changed, and bad news has kept on coming, alongside increases in greenhouse gases.

Communication is now more direct because, at the COP meetings in Copenhagen and Paris show, the discussion is: this is going to arrive and now it's here. The rules are changing, it's tangible; people are suffering because of it right now. Even Greta Thunberg's language has changed between her Davos speech, her UN speech and finally her COP 25 speech in Madrid. It's becoming more and more aggressive with a final demand for changes in our ways of acting, with communications that could even be compared to those of Martin Luther King: the worst thing isn't inaction, it's people who are pretending to act and doing nothing.

For all that, isn't there still a place for positive communication, especially at the local level?

It's not antithetical. All 'corporate' discourse can be positive by insisting on the fact that progress is being made. It isn't because you have a grassroots method of communication that's becoming radicalized that all ways of communicating will change. That being the case, all discussions are cranking things up a notch in terms of intensity, demands or proof of action, but public and private actors, as well as institutional ones, will be using positive communications by saying that we are in a transition period. However, in two years, the perception of the threat has markedly developed to become a greater threat that is reflected in presses such as *The Guardian*, for example, that have decided to no longer talk about global warming in a way that has positive connotations and instead refers to climate catastrophes or dangers.

We have moved on from a largely scientific discussion that integrates the public space to a more radical discussion that integrates the political sphere. We have very quickly gone from citizens being in denial to them challenging the actions of politicians against a backdrop of collapse. There is therefore a general semantic revision that is creating tensions, and public and private scientific discussions are currently very challenged by the development of conspiracy theories; politicians are aware but they aren't doing anything.

In such conditions, how should we talk to children?

It's very complicated. For a long time, we were set on a very educational communication along the lines of 'this is happening because of greenhouse gases'. But children, even at a very young age, are becoming very radical because at school they talk about it in a very aggressive manner with very negative language that is taken up in the public space. So, if we use very positive language to explain to them everything that's happening, there is the risk that they'll perceive it as a lie. This creates stress and anxiety around the idea of the end of the world that is popping up in the schoolyard, even if it is perhaps quite a Parisian phenomenon. It's suddenly very complicated to find a suitable register for communication.

Lastly, should communication be focused on changing behavior or on accepting more restrictive public policies?

There is currently a significant debate around the issue of metanarratives to lead people towards a new vision of the world. However, the world is polyphonic and so it's difficult to want to impose a new emancipatory discourse when everyone is altering its meaning themselves. Regarding solutions, young people have already began using very restrictive, 'survival mode' solutions, while older people lean more towards 'nudges', little changes. It's relatively certain that 30–50 year olds will change their behaviors but that they will have limited exposure to the consequences of climate change. Someone under 30 knows that they will be much more exposed and will therefore be ready to abandon their rights and freedoms and accept more restrictive standards and regulations.

Box 8.1. *Interview with Jacques-Olivier Barthes, Director of Communication of WWF France*

8.5. Conclusion

The research carried out on the subject shows that communication about climate change that is too alarming is generally rejected by consumers, whether the threat is perceived as exaggerated in relation to consumer skepticism, or whether the consumer has the impression that there is little to be done at the individual level, due to the scope of the phenomenon, which leads to the establishment of individual defense mechanisms. These elements make us think that it is necessary to rethink communication strategies on climate change in order to restore confidence in our individual and collective abilities, in order to act in an effective manner. However, today we still talk in terms of mass extinction, alarmist visions that are without a doubt more interesting from the point of view of the media but that can have devastating effects on individual engagement. Documentaries and communications campaigns emphasizing simple solutions that are able to be integrated into daily routines should therefore be proposed, while also explaining the concrete effects that such actions can have on climate change.

Moreover, as consensus-based as it is, it seems that Witte's EPPM has two limitations. Firstly, it does not take into account the issue of psychological distance from risks, which suggests that the image of a close risk would have the same persuasive impact as an image of distant risk. However, a number of works have shown that, depending on the perceived distance of the risk, the individual will have neither the same interest in the threat nor the same will to combat climate change (see McDonald *et al.* (2015) for a review). Secondly, the EPPM does not take into account the degree of abstraction of the proposed response. In other words, it does not consider the fact that the proposed solution may be described in an abstract manner or in a concrete manner, which could have an impact on how individuals perceive its efficacy (see Han *et al.* (2016) for a review). These two dimensions are excluded from the EPPM, even though they could play a decisive role in activating one of the processes mentioned, that is, the danger-control process or the fear-control process. This gives us two other areas for improving communication on climate change. The first consists of bringing the threat closer to ensure that people are worried about it and cannot fall into a strategy of denial or rejection, and the second consists of working on solutions to make them more concrete.

At this time, when we are experiencing an unprecedented mobilization of young people, the specific case of children is interesting to study. While the school is a significant agent of socialization for young people and it implements educational actions and programs in accordance with the issue, it is nonetheless also necessary to consider the manner of promoting engagement within families and couples. This subject has not been widely addressed in the literature and deserves further consideration to enable young people to have the best cognitive, emotional and behavioral education regarding climate change. Involving all of the agents of socialization of young people would make it possible to make the issue more concrete in their eyes and promote their engagement. To conclude, it appears to be important to provide the communication best suited to the young generation, in line with their age and concerns. However, future work on this issue must also consider the voice of the young generation and must take it into account decision-making regarding climate change.

8.6. References

Akil, H., Robert-Demontrond, P., Bouillé, J. (2018). Exploitation of mortality salience in communication on climate change. *Recherche et applications en marketing*, 33(1), 2–29.

Baumeister, R.F., Vohs, K.D., Nathan DeWall, C., Zhang, L. (2007). How emotion shapes behavior: Feedback, anticipation, and reflection, rather than direct causation. *Personality and Social Psychology Review*, 11(2), 167–203.

Dillahunt, T., Lyra, O., Barreto, M.L., Karapanos, E. (2017). Reducing children's psychological distance from climate change via eco-feedback technologies. *International Journal of Child-Computer Interaction*, 13, 19–28.

Feinberg, M. and Willer, R. (2011). Apocalypse soon? Dire messages reduce belief in global warming by contradicting just-world beliefs. *Psychological Science*, 22(1), 34–38.

Han, D., Duhachek, A., Agrawal, N. (2016). Coping and construal level matching drives health message effectiveness via response efficacy or self-efficacy enhancement. *Journal of Consumer Research*, 43(3), 429–447.

Haynes, K. and Tanner, T.M. (2015). Empowering young people and strengthening resilience: Youth-centred participatory video as a tool for climate change adaptation and disaster risk reduction. *Children's Geographies*, 13(3), 357–371.

Klein, N. (2019). *On Fire: The (Burning) Case for a Green New Deal*. Simon & Schuster, London.

Maibach, E.W., Roser-Renouf, C., Leiserowitz, A. (2008). Communication and marketing as climate change-intervention assets: A public health perspective. *American Journal of Preventive Medicine*, 35(5), 488–500.

McCright, A.M. and Dunlap, R.E. (2010). Anti-reflexivity. *Theory, Culture & Society*, 27(2/3), 100–133.

McDonald, R.I., Chai, H.Y., Newell, B.R. (2015). Personal experience and the "psychological distance" of climate change: An integrative review. *Journal of Environmental Psychology*, 44, 109–118.

O'Neill, S. and Nicholson-Cole, S. (2009). "Fear won't do it": Promoting positive engagement with climate change through visual and iconic representations. *Science Communication*, 30(3), 355–379.

Ojala, M. (2012). Regulating worry, promoting hope: How do children, adolescents, and young adults cope with climate change? *International Journal of Environmental & Science Education*, 7(4), 537–561.

Schill, M. and Ezan, P. (2018). Les représentations sociales du réchauffement climatique par les enfants de 7 à 12 ans. In *34ème Congrès international de l'Association française du marketing*, Strasbourg.

Schill, M. and Odou, P. (2017). Réactions affectives anticipées et intentions d'agir contre le réchauffement climatique : proposition d'une typologie. *Décisions Marketing*, 88, 13–33.

Walker, C. (2017). Tomorrow's leaders and today's agents of change? Children, sustainability education and environmental governance. *Children & Society*, 31(1), 72–83.

Witte, K. (1992). Putting the fear back into fear appeals: The extended parallel process model. *Communications Monographs*, 59(4), 329–349.

Witte, K. and Allen, M. (2000). A meta-analysis of fear appeals: Implications for effective public health campaigns. *Health Education & Behavior*, 27(5), 591–615.

9

Environmental Regulations and Awareness-raising Campaigns: Promoting Behavioral Change through Government Interventions

Leila ELGAAIED-GAMBIER and Laurent BERTRANDIAS
Toulouse Business School, France

9.1. Introduction

In 2019, the French government was sued for "climate inaction". Known as "the case of the century", this initiative was led by four NGOs, including the Nicolas Hulot Foundation, which was behind the petition that received over 2 million signatures. Similar intiatives were seen in other countries (for example, the Netherlands, the United States), reflecting the high expectations civil society has of governments in terms of climate change. In 2020, the (in)action of political decision-makers is still at the heart of many mobilizations. It currently seems that the question is no longer about whether states should intervene, but how they should intervene to reduce the effects of climate change.

The French Environment and Energy Management Agency (ADEME) regularly publishes calls for projects intended to create a dialog between researchers and

Marketing for Sustainable Development,
coordinated by Sihem DEKHILI. © ISTE Ltd 2021.

decision-makers. The aim is clear: "develop recommendations for public policies".[1] Researchers are invited to present potential interventions in support of sustainable development, while also describing how such measures would be operationalized. While this willingness to get researchers to participate in considerations is a positive development, it also highlights how difficult it is for the public authorities to suggest relevant environmental policies. The scale of the challenge is significant as governments need to design environmental transition measures that address a diversity of imperatives including efficacy, feasibility, costs, acceptability to different stakeholders, equity between citizens, etc.

Public-policy makers have significant means at their disposal to bring about behavioral change among both consumers and companies. For William Nordhaus, winner of the Nobel Prize in Economics in 2018, the solution to limit the environmental footprint of our activities would be to pay a price for carbon emissions. Such a carbon tax would apply equally to governments, companies and households. However, in terms of environmental taxation, the transition from theory to practice has been complicated. Recently, in France, the announcement of a tax on diesel triggered demonstrations and was the spark that led to the emergence of the "*gilets jaunes*" ("yellow vests") movement. In 2016, the "ecotax" was permanently abandoned under pressure from a coalition defending regional interests. However, its environmental benefits were almost universally recognized.

This raises a number of questions: how can governments make citizens adhere to environmental measures that require them to sacrifice a little of their well-being today to significantly improve the well-being of future generations (probably including their own descendants)? How can taxes and restrictions to individual freedoms for the benefit of the public good and general interest be made to be accepted? How can behavioral changes that require a loss of comfort, or even a slight financial loss, be encouraged? How can businesses be convinced when their market position risks being challenged? As complex as they are, these questions strike directly at the ability of the public authorities to justify pro-environmental measures in the eyes of the actors affected by such regulations.

Marketing can be seen as a means of legitimization (Humphreys 2010). This chapter is intended to show how the consideration of marketing in public policies, in support of sustainable development, can help to make those policies more acceptable and more effective.

1 See: https://seminaire-transitions-ecologiques-economiques-sociales.site.ademe.fr/.

After detailing the different types of environmental regulations and their challenges, we will present areas for reflection and discuss the potential contributions marketing can make towards strengthening the efficacy of pro-environmental public policies.

9.2. Overview of the environmental intervention tools of public authorities

In terms of environmental regulations, public authorities have four main and very different approaches and rely on distinct mechanisms for behavioral change: coercive measures, incentivizing measures, communication and nudges. We will now detail each of these regulatory approaches.

9.2.1. *Coercive environmental measures: the most radical approach*

The first option for governments is to adopt compulsory measures to reduce the environmental footprint of individuals and organizations through the enforcement of laws. Coercive measures involve sanctions in the event of non-compliance, most often in the form of fines. For example, the French government established an air quality certification called Crit'Air[2], which aimed to reduce air pollution by moving the most polluting vehicles out of towns in favor of vehicles with lower emissions. Motorists obtained a sticker indicating their vehicle's environmental classification in line with its emissions. There are six categories, some of which can be prohibited from driving within restricted traffic areas or at peak times for pollution. In cases where there is a violation of traffic restrictions or the absence of a sticker, the transgressor is subject to a fourth-class fine for heavy trucks, buses and coaches and a third-class fine for other vehicles, equal to 135 or 68 euros respectively. While the anticipated effects of this measure would be in the general interest (less pollution and thus fewer negative impacts on health and the environment), it seemed *a priori* unpopular among French people. We can suppose that its main detractors are precisely those who will be negatively affected by this traffic ban. Individuals do indeed identify the loss of personal freedom as the main source of dissatisfaction in relation to strict regulations (Attari *et al.* 2009).

Whether it is a matter of environmental preservation or public health, public authorities regularly encounter the question of whether it is preferable to convince the public or impose more virtuous behavior on them. While the second option is

2 See: https://www.ecologique-solidaire.gouv.fr/certificats-qualite-lair-critair#e5.

more radical, it has sometimes been necessary, especially if the urgency of the situation no longer makes it possible to take time to change beliefs through persuasion. However, previous research seems to go in the direction of persuasion, rather than coercion. Using self-determination theory as support, Moller *et al.* (2006) show that public policies encouraging self-regulation and independent decision-making are more effective means of encouraging sustainable behavior than forced adoption. Prior work highlights the crucial role of intrinsic motivations and autonomy in the adoption of a range of behaviors, including pro-environment behaviors (for example Pelletier *et al.* (1998)). If we take the example of regulation through the Crit'Air label, motivations to comply with the traffic ban will be linked to a genuine belief in the usefulness of the measure and a desire to reduce pollution (intrinsic motivation), or a desire to avoid the potential sanctions, namely paying a fine of 68 euros (extrinsic motivation).

It is thus worth investigating individual consent to comply with such measures when intrinsic motivations are non-existent (for example, a lack of involvement, questioning the merits of the measure), or extrinsic motivations are insufficient (for example, insufficiently dissuasive sanctions, lack of gratification).

9.2.2. *Ecotaxes and financial incentives: taxation as a dissuasion or an incentive*

The second tool to promote behavioral change is taxation. The underlying idea is to apply the "polluter pays" principle. In short, it involves discouraging bad practices and encouraging good habits by altering the amount owed in line with actual behavior. In this context, taxation can be used in either an incentivizing or punitive manner.

Incentivizing taxation consists of encouraging private individuals or companies to act in the public interest in exchange for tax benefits (for example, tax credits, lower taxes, tax exemptions, reduced license fees, etc.). In this case, taxation acts as a reward. For example, in regard to waste processing, local authorities have the option to establish a tax incentive that determines the amount paid by users in line with the volume of waste they produce.

In other cases, taxation is instead considered a deterrent. Bad practices are thereby punished by means of specific taxes or an increase in taxes. Dissuasive taxation is often seen as unfair and public authorities are struggling to justify it in order to establish it. Its rejection is often encouraged through the use of archetypal figures, i.e. symbolic images acting as universal models, and these make it possible to destabilize the tax implementation process: for example, the driver living far from the city center who has to drive a long way to get to work, or the head of a small or medium-sized business crushed by restrictions and taxes.

9.2.3. *Environmental information, awareness-raising campaigns and persuasion: the crucial role of education*

The third approach is to raise awareness and convince people to adopt sustainable practices without resorting to coercion or the reward/punishment pathway. This approach preserves citizens' freedom to choose and relies on an internalization of pro-environmental standards, which leads individuals to act on the basis of more intrinsic motivations. However, this also requires significant powers of persuasion. From this perspective, communications and advertising have an essential role to play. Much research has highlighted the key role of environmental information in the adoption of environmentally friendly behaviors. In particular, environmental labelling and Ecolabels can be a good means of providing information about the carbon footprints of some products, raising awareness or convincing consumers of the importance of behavioral change (Vanclay *et al.* 2011). The efficacy of this method remains limited as such labeling is not systematic (Bernard *et al.* 2015).

Public authorities have failed to make widespread environmental labelling mandatory, despite the fact that this is one of the main measures noted in the Grenelle Environment debates in France. The project did not lack support, but was hindered by the effective opposition of professional organizations who knew how to destabilize the institutionalization process (François-Lecompte *et al.* 2017). As a result, similarly to taxation, the implementation of informative mechanisms to raise awareness relies on first convincing a range of stakeholders to collaborate.

Using advertising as a means of persuasion is sometimes presented as a credible means of avoiding strict regulations, while achieving behavioral change voluntarily, rather than by force. However, a relatively small number of academic research projects have analyzed the efficacy of pro-environmental government communications. Moreover, the communications of the public authorities have long been subject to criticism on a number of fronts, with some questioning the use of taxpayer money, in particular (Mullen and Bowers 1979).

9.2.4. *Green nudges: using behavioral science to serve environmental public policies*

Lastly, pro-environmental behaviors at the individual level can be encouraged through green "nudges".

Nudges do not rely on persuasive mechanisms, incentives or sanctions. A nudge refers to "any aspect of the architecture of choice that changes the behavior of individuals in a foreseeable manner, without prohibiting any option, or altering their economic incentives in any significant way" (Thaler and Sunstein 2008, p. 6). With that in mind, green nudges consist of pushing individuals towards adopting environmentally friendly behaviors by simply changing the decision-making context, or by presenting information in such a way as to subtly encourage them. Schubert (2017) identifies three types of pro-environmental nudges.

The first category uses highlighting mechanisms. The aim of a nudge is to draw attention to certain features of a product, or to an aspect specific to the context in which the action takes place. Recurring examples include stickers being placed on switches to remind people to turn off the lights when they leave a room, or even putting the outlines of footprints on the ground to lead the way to recycling bins, in order to make them more visible. Ecolabels and environmental information placed on products at the point of sale can also be considered forms of nudges.

The second category of green nudges involves the use of social norms. For example, some electricity providers have chosen to compare the energy consumption of individual households with that of their neighbors and regularly notify them of their respective changes. Another well-known example is that of messages in hotel rooms encouraging the reuse of towels, making use of a descriptive norm stating that the majority of guests staying in the same room have reused their towels with the aim of reducing their environmental footprint. It can be difficult to differentiate this type of nudge from traditional communications. Indeed, persuasive communications can also make use of social norms to convince people. In theory, to be considered a nudge, the mechanism must be used while the behavior is taking place, i.e. it must target the moment of action (for example, the precise moment when someone leaves a room, the moment they print something, the moment they are about to throw away a cigarette butt, etc.). While communication often targets attitude change and can raise awareness at any time, nudges are solely focused on behavior change.

Lastly, the final category of green nudges is based on the use of default options. This involves the preselection of choices, while leaving the individual free to choose an alternative solution. For example, an electricity provider can automatically sign its customers up to a renewable energy program with the possibility of easily leaving this program. A company can even make the option of double-sided printing the default, in order to save paper.

The efficacy of these helping hands lies in the tendency of individuals to look for a positive self-image, conform to social norms or follow the movement by restricting themselves to pre-selected choices (Schubert 2017). Nudges can indeed be effective in triggering behavioral change without coercion, but the new behavior does not necessarily rely on convictions or on a change in attitude. While it is used increasingly widely in government policies the world over, the nudge theory has been subject to significant criticism. Manipulation (is the underlying intention always clear and transparent to the target audience?) and paternalism (who can decide what is really in the personal interest of an individual?) are central to discussions. More generally, nudges are effective if they are designed for a specific context, which makes their adoption largely dependent on the interest of local decision-makers in using such measures.

9.2.5. *Towards an optimal regulatory mix*

Public authorities thus have various complementary regulatory tools (see Figure 9.1). Each of the tools mentioned above have their own advantages, as well as some limitations (see overview in Table 9.1). In order to obtain effective results, they should be combined by being used simultaneously, separately or in turn, depending on their target audience (citizens or companies), the type of behavior targeted, the maturity of public opinion and so on. Box 9.1, which explains the work carried out by water agencies, is a good illustration of this need to combine coercive, financial and awareness-raising measures. This combination of tools is contributing to the emergence of a kind of regulatory mix, along the lines of the marketing mix, or the communications mix. These instruments should therefore not be considered in isolation, but as cogs within a complex mechanism. Following the example of the elements of the marketing mix, the different green policies must be used to further a comprehensive environmental strategy for public authorities. This inclusive vision makes it possible to leave behind a strictly regulatory logic to adopt a logic based on objectives. This also makes it possible to avoid potential conflicts or contradictions that could occur among environmental measures; they are considered in a consistent and harmonious manner to serve one and the same purpose. The mixture of the different measures thus stems from the objectives in question, which must be explicitly established in advance.

Type of environmental regulation	Advantages	Limitations
Coercive environmental measures	– In theory, quick results, very high level of effectiveness	– Unpopular, often perceived as unfair – Loss of free will of actors – Difficulty of establishing suitable sanctions (appropriate but sufficiently dissuasive) – Behavioral change is not necessarily accompanied by a change in attitude – Risk of civil disobedience
Ecotaxes	– Makes it possible to free up resources for actions in support of sustainable development	– Behavioral change relying on extrinsic, and thus less effective, motivations – Unpopularity of taxes as a deterrent (being fed up of taxes) – Lack of fairness, with the negative impacts of ecotaxes generally more pronounced for low-income households
Tax incentives	– Possibility of adopting a reward framework or a sanction framework to optimize effectiveness – Makes it possible to begin the behavioral change process	– Can prove costly and difficult to sustain – Risk of a 'deadweight' effect in the case of incentivizing measures – Risk of destabilizing sectors of the economy when the mechanism is challenged
Awareness-raising and persuasion	– The free will of citizens is preserved – Behavioral change is generally accompanied by a change in attitude and the internalization of pro-environmental standards	– Requires a lot of time and repeated efforts – High cost – Limited efficacy – Being informed and developing a positive attitude towards environmental action does not mean that behavior will follow suit (the attitude/behavior gap)
Green nudges	– Inexpensive – No sanctions or incentives – The free will of citizens is preserved in theory (they can choose to ignore the nudge) – In some cases, quick and tangible efficacy	– Behavioral change is not necessarily accompanied by a change in attitude – Temporary efficacy – Ethical debate: often integrated into manipulation and some paternalism – Need to refresh nudges regularly and adapt them to context

Table 9.1. *Overview of advantages and limitations of pro-environmental regulatory measures*

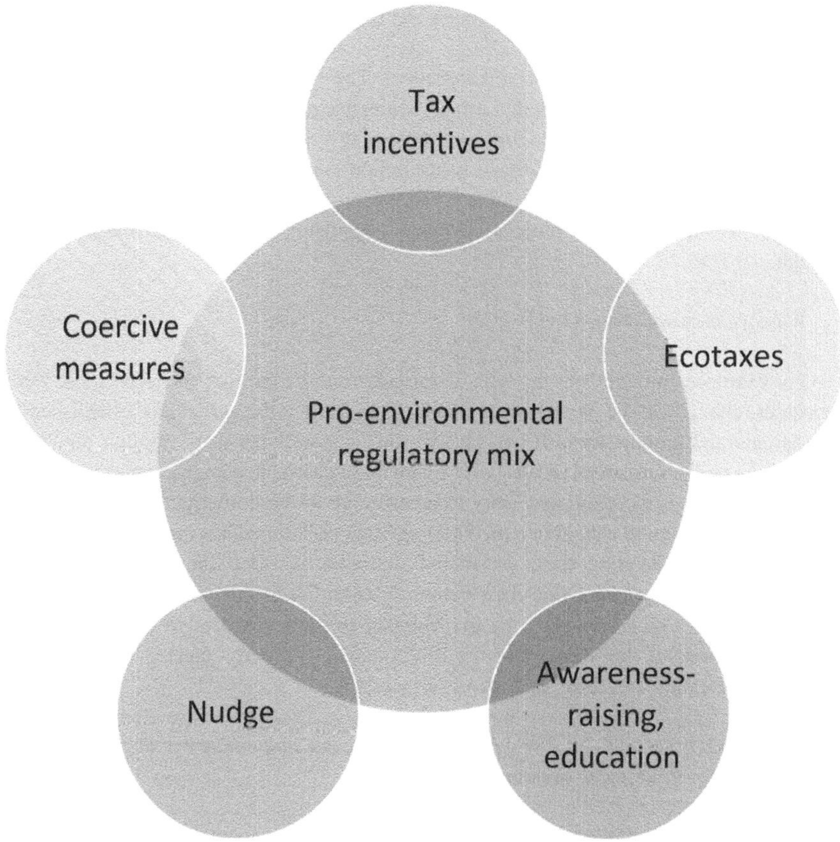

Figure 9.1. *Proposal for a regulatory mix combining the different types of environmental measures to be used by the public authorities. For a color version of this figure, see www.iste.co.uk/dekhili/sustainable.zip*

What is the Water Agency and what are its aims?

The Water Agency is a public body under the joint auspices of the Ministry of the Economy and Finance and the Ministry of the Environment. There are six Water Agencies in France, organized by drainage basin. These are financially independent bodies thanks to the collection of charges on water bills, which are then reinvested into projects to improve water quality. Our intermediaries are largely local governments, as well as all local actors involved in water management. A Water Agency has various aims, including guaranteeing the conservation of water resources, combating pollution, protecting marine environments and biodiversity, and helping project managers to optimize water management.

What types of levers are used to promote the responsible management of water?

First of all, there is the coercive component. The regulatory framework is the water framework directive, which has the aim of achieving good water status. This therefore involves helping with all projects that contribute to this aim. All of the actions carried out must also be approved by the national water police. Sanctions are provided for cases of violation of the regulations. Then, there is the incentivizing aspect. We propose financial assistance to support certain projects. However, regulatory and financial means are not enough.

Why are they not enough?

For example, in regards to the type of agriculture to be put in place to preserve water resources, change can be perceived as a lack of respect for the agriculture established by our parents and grandparents. It can be very difficult to reach them through regulatory discussions or even financial means, with financial sanctions low in relation to investment in compliance. It is therefore necessary to be attentive to these intermediaries, in order to try and support them in changing a methodology that can sometimes involve an emotional dimension. Understanding these emotional obstacles is often the first step towards convincing people and changing their beliefs. In addition to this, public policies can be contradictory; the measures taken by the Ministry of Agriculture often diverge from the recommendations of the Ministry of the Environment. Similarly, elected representatives often have economic interests, such as preserving jobs.

What strategies can help to strengthen the legitimacy and efficacy of the measures put in place for improved water management?

Water Agencies don't usually communicate with the general public and prefer to focus on work with project managers. However, in recent years, we have tried to expand our communications with more accessible campaigns. What helps us today is the existence of a social movement, pressure from consumers and citizens who are demanding a different agricultural practice and who need to know the story behind a product. Relying on other actors allows us to change opinions bit by bit.

We also propose establishing educational workshops for elected representatives, as they remain the principal decision-makers. This would provide them with general information about the water situation, familiarize them with the concepts of small and large water cycles, and present them with examples of projects that have been carried out to normalize good practices.

We are currently involved in a logic of reasoning with the different actors, and we are trying to find a shared solution instead of imposing our vision.

Box 9.1. *Interview with Jade Pellerin, Operations Officer at the Seine-Normandie Water Agency*

9.3. Improving the effectiveness of pro-environmental public policies: the contribution of marketing

Climate change and environmental damage require decision-makers to implement increasingly effective actions to change the behavior of actors. The efficacy of policies is often harmed by the resistance of these actors. As a result, many measures are ultimately adopted and put in place in a reduced form, and their overall impact is therefore limited. Sometimes they are even abandoned, thereby generating irrecoverable costs that taxpayers find difficult to accept.

Pro-environmental measures and regulations are likely to have an impact on a large number of actors capable of exercising resistance. For example, the draft directive prohibiting single-use plastic has been postponed to 2040, owing to economic arguments put forward by the sector. Even if reducing the environmental impact is obviously legitimate, such resistance translates into the existence of competing legitimacies, and one of the major challenges of the marketing approach is to realign these legitimacies. The "megamarketing" approach and the actions involved could contribute to this.

9.3.1. *Adopting a megamarketing approach to increase the chances of success of pro-environmental measures*

Kotler (1986) proposed the concept of megamarketing to reflect the idea that, in many situations, client-oriented marketing is not enough. The success or failure of a company often puts political aspects at stake, which requires a departure from the strict customer relations framework.

Beyond the target customers, a company must therefore pursue a "strategically coordinated" effort "to obtain the cooperation of all stakeholders, with the aim of entering or operating in a given market" (Kotler 1986, p. 117 118). In concrete terms, this means that the company must "create incentives or exert pressure to the right extent and at the right moment" (Kotler 1986, p. 117), in order to secure the collaboration of actors who, without being direct customers, are nonetheless essential to the success of the company. For Kotler, megamarketing is mainly necessary for penetrating markets or avoiding being shut out of them. Humphreys (2010) shows that megamarketing primarily involves institutional work, which must serve to legitimize a new practice or activity when it comes up against barriers linked to regulation, existing norms or established beliefs.

Surprisingly, the concept of megamarketing has not been widely used to analyze the success and failure of public policies. In contrast, this frame of reference is acknowledged to be particularly useful in cases where a common practice is challenged. Pro-environmental regulations enter into this situation to the extent that they are intended to change the behavior of the actors. The process of institutionalizing widespread environmental advertising promoted by the government following the Grenelle Environment debates is an illustration of the challenge of megamarketing. Supported by pioneering companies, the government wanted companies to provide systematic information on the environmental impact of products. However, the efforts made were blocked by opponents working to hinder the implementation of the mechanism. This work, mainly carried out by trade unions in certain sectors, consisted of challenging the cognitive basis of widespread advertising, particularly by highlighting the unsustainable costs of implementation, the complexity, the unfinished nature of the methodology and even the difficulty of monitoring labelling on imported products (François-Lecompte *et al.* 2017). The megamarketing approach provides opportunities for managing these types of situations by reconciling marketing and political strategies. This is based on the development of economic, political, psychological and interpersonal skills. In the following sections, we will discuss the aspects that seem key to pro-environmental measures.

9.3.2. *Identifying competing legitimacies and mapping power structures*

As Kotler advises, an initial challenge in facilitating adoption is correctly identifying the competing legitimacies and assessing the degree of power of the stakeholders who wield it. This must enable the development and appropriate incentives and/or pressures. This essential task is also one of the most difficult as, in the case of pro-environmental measures, the power structures do not necessarily overlap with structured bodies. For example, the "*Gilets jaunes*" movement was born because the increase in taxes on diesel was considered unlawful. However, the scale of movement has been a surprise, as has the fact that it is not directly linked to an existing political or union-related movement. By drawing on works of political science, Kotler draws up a list of three types of power structure, all of which are relevant in the case of pro-environmental measures:

– *pyramid structures*: in which the power lies with a ruling elite, whose support must be obtained. In the case of pro-environmental approaches, this may involve an industrial group or sector of activity capable of effectively hindering the institutionalization of pro-environmental measures. Some examples could be airlines and airplane constructors, who regularly oppose the establishment of an aviation tax. Capable of powerful lobbying, they emphasize, for instance, their actions in support of the environment (carbon offsetting through reforestation, R&D aimed at the development of efficient or electric engines, etc.). This allows them to discredit the tax or, at least, put its use into perspective;

– *factional structures*: in which pressure groups and interest groups fight to gain power. Political parties can play a role in supporting or combating pro-environmental regulations, depending on whether or not this helps them to enter into power;

– *the power of coalitions*: through which different blocs will temporarily become allies to achieve shared goals. The “*Bonnets rouge*” (“Red caps”) movement, created in France in response to the implementation of the Ecotax and aiming to bring about its revocation, is a textbook case as it brought together business leaders, unions of agricultural workers, unionized industrial workers and local elected representatives.

While the identification of power structures, their relative influence and the legitimacy of their aims in the eyes of the general public is complicated, it is absolutely necessary to be able to take effective action. Kotler’s recommendations relate solely to political strategy and the author thus suggests implementing actions to neutralize the opposition. For example, through the establishment of a methanation plant, ADEME encourages the proponents of the project to ensure that the structure is well integrated into its environment and to reduce potential pollution as far as possible. The aim is to mitigate or reduce the resistance of residents. Kotler also suggests building coalitions of supporters and even turning neutral actors into allies to increase the chances. The awareness-raising work carried out recently by ADEME among municipal election candidates[3], in order to guide them through the green transition once they are elected, is a good illustration of this. As important as political strategy is, the analysis by Humphreys (2010) allows us to broaden our considerations and encompass typical marketing practices. By relying on the new institutionalism theory, the author insists on the need to establish new cognitive patterns, and change normative structures (what is and is not desirable in society) and regulatory structures. The Water Agency, which must work on changing the beliefs of farmers, in order to lead them towards developing more sustainable practices, is a good example of this (see Box 9.1).

9.3.3. *Understanding the cognitive patterns of individuals*

In order to be legitimate, pro-environmental measures must be based on institutional work, creating new routines (Ben Slimane and Lecas 2010). In order to achieve this, it is important, first of all, to develop a better understanding of the

3 See: http://www.courrierdesmaires.fr/84401/elections-municipales-lademe-veut-convaincre-les-futurs-maires-hussards-de-la-transition-ecologique/.

psychology of the individual. An initial stage consists of studying the cognitive[4] and heuristic[5] biases linked to environmental issues. In some cases, these heuristics prevent people from correctly interpreting environmental information. For example, studies suggest that individuals would have a poor understanding of climate change, owing to a mental framework pushing them towards understanding it as a static object and not as a dynamic process, thereby creating inertia at the behavioral level (see, for example, Chen (2011)).

Understanding the psychology of individuals makes it possible to adjust regulatory measures in a way that will create compliance and reduce sources of tension. Some promising measures can, in fact, prove to be counterproductive if they do not take personal mental mechanisms into account.

For example, programs with monetary incentives for household waste can produce very different outcomes depending on their guidelines (Tversky and Khaneman 1981). The monetary incentive can be presented as a financial gain (a reward in the form of lower fees) if the consumer presents their trash less often than the maximum number of collections organized per year, or as a financial loss (a punishment in the form of a fee increase) if the consumer presents their trash more often than a threshold fixed by the union of waste collectors and processors. While the two systems are economically equivalent, highlighting different salient points can lead to varied, and even opposing, reactions. Overall, the academic literature suggests that the "punishment" approach would be more effective, particularly because of an aversion to loss (Tversky and Khaneman 1981): the idea of losing money is less acceptable than a renunciation to earn it.

However, the "reward" approach is inherently positive. Throwing away less is an act that can be acknowledged monetarily. The "punishment" approach deprives users of this type of gratification and relegates them to the less than encouraging status of polluter. This can also trigger psychological reactance mechanisms as it restricts freedom. The user may thus ignore the message, continue to throw things away, stop paying more (this is all the more conceivable because financial incentives are generally modest) or even adopt deviant behaviors (filling up other people's trash cans, littering, etc.).

---

4 A cognitive bias is a systematic distortion of judgment that prevents a rational processing of information. This is caused by the tendency of the human brain to perceive information through a personal filter (past experience, preferences, etc.).

5 Heuristics refers to the automatic mental processes used by individuals to form judgments and make decisions in a fast and intuitive manner. These cognitive shortcuts make it possible to simplify the decision-making process and save time, but, in some cases, they can lead to errors in judgment.

Considering these psychological mechanisms before making regulatory decisions is proving to be essential. However, citizens are not a homogenous mass and some individual differences deserve to be considered by regulators, particularly through segmentation.

9.3.4. *Segmenting the "market" to optimize legitimization strategies*

While the input of marketing is acknowledged as being useful within the framework of establishing and promoting public action, some key concepts of the field are set aside, owing to their perceived contradiction with the requirements of the public administration. Segmentation, in particular, a fundamental stage in any marketing strategy, would be incompatible with the universality approach of public services. According to the dominant perception:

> Taking into account the heterogeneity of the population through segmentation is limited within public administration, as all citizens must be treated equally by public services. The private sector presupposes that it is not possible to serve everyone in the same manner, which is not applicable in the public administration. (Kaplan 2007, p. 27)

However, segmentation does not necessarily involve treating citizens in an unequal manner. It can simply mean taking their differences into account in order to serve them better. For example, for the collection of household waste, some local authorities that have put a financial incentive in place are trying to make the mechanism fairer by making different size receptacles available. The size of the receptacle provided generally depends on the number of people in the household. As the incentive is linked to the number of times the receptacle is presented for collection during a year, this type of segmentation by household size contributes to the acceptability of the financial incentive to everyone and makes it more effective (Figure 9.2).

Furthermore, in order to raise awareness of certain environmental issues and change behaviors, state institutions should not necessarily address all citizens in the same way. Proper segmentation makes it possible to target people better and *in fine* to optimize strategic efforts and the associated costs. If one part of the population has specific needs or limitations, there is little point in persisting in using the same incentives as for the rest of the population. For example, some people are required to make significant use of their personal vehicle, either because the public transport network is not well developed in their area, or because their professional activity demands it. It is not relevant to target these individuals to promote the use of public transport.

Moreover, following the innovation adoption cycle, some citizens act as early adopters or opinion leaders that can pass on good practices and participate progressively in normalizing new behaviors. The latter can also constitute a first stage towards making some environmental measures more widespread.

Door-to-door collection: the volume of the trash can

You may choose the size of your receptacle. This depends on the number of people in your household:

- 1 person: 80 liters
- 2/3 people: 120 liters
- 4/5 people: 180 liters
- 6/7 people: 240 liters
- 8/9/10 people: 360 liters
- Professionals: free choice, including a dumpster (660 liters)

Figure 9.2. *Example of segmentation by household size as part of establishing a financial incentive for household waste (source: website of the Association of Municipalities for the Collection and Processing of Household Waste in the Châteauneuf-sur-Loire Region[6])*

9.3.5. *Establishing legitimization strategies: the crucial role of communication and education*

The support of citizens is critical to the success of government policies. Box 9.1 shows that the development of social expectations in regard to the agricultural model, for example in terms of the use of pesticides, helps water agencies to carry out projects to protect the resource. The delicate question of the compliance of citizens with environmental legislation has been of increasing interest to researchers in recent years (Rhodes *et al.* 2017). However, in order to ensure the compliance and cooperation of individuals, it is necessary to expose them to the followers and outcomes of regulatory decisions. Under the new institutionalism theory, legitimacy has three dimensions, regulatory, normative and cultural-cognitive (Suchman 1995), which are supposed to complement and strengthen each other. Regulatory legitimacy refers to the capacity to impose through rules, as well as the capacity to ensure compliance with the process of creating rules. In this way, all environmental measures must be supported by education and prepared in advance. This is all the

6 See : www.sictom-chateauneuf.fr.

more true if it naturally creates reluctance among citizens or other stakeholders and disrupts its implementation. The establishment of mandatory measures and ecotaxes, in particular, requires significant educational efforts before, during and after the implementation of the measure. That being said, it is risky to restrict institutional work to the regulatory level as public authorities find it difficult to be effective at the normative and cultural-cognitive levels: a pro-environmental measure may be well accepted as a concept, without producing concrete impacts. The first stage, which consists of providing information, is therefore indispensable but insufficient. So how can the other two pillars be used?

Figure 9.3. *Example of communication by ADEME (the French Environment and Energy Management Agency) on Twitter to establish new cognitive patterns, following a measure on the progressive prohibition of single-use plastics. For a color version of this figure, see www.iste.co.uk/dekhili/sustainable.zip*

Regarding the cultural-cognitive aspect, one of the major underlying challenges in the adoption of ethical environmental behaviors is establishing or re-establishing forgotten cognitive patterns. For example, reclaiming or repairing objects and choosing the sustainable over the disposable are no longer simple. In many ways, companies have encouraged rapid replacement and the choice of easily disposable or single-use products. As a result, in order to be legitimate, pro-environmental measures must be based on institutional work consisting of creating new routines and new practices, and establishing new standards (Ben Slimane and Lecas 2010). Public authorities cannot do this alone and must rely on intermediaries that are able to establish cognitive patterns. For example, ADEME publish guides on responsible consumption, are very active on social media, and finance research that makes it possible to strengthen the cognitive legitimacy of a given change (Figure 9.3).

Establishing new patterns and beliefs is essential. However, while pro-environmental measures are intended to benefit new practices, these are only really adopted when they are normalized, i.e. considered desirable in a given social context. The work of normative legitimization can thus entail communication based on the activation of social norms in favor of protecting the environment. Social influence theories highlight our tendency to conform to the norms dictated by groups to which we belong or refer. The difficulty lies in mobilizing this lever of influence when the environmental behaviors that we seek to encourage are simply not the norm, when they are not sufficiently widespread. In such cases, it is not possible to communicate by emphasizing a mainstream tendency. Let us take the example of online green behaviors. At present, only a small minority of the population is aware of the environmental impact of their digital behavior and try to remedy it by using green search engines, limiting streaming and saving files in the cloud, or even by cleaning out their inboxes frequently. As such environmentally friendly digital practices have not yet been widely adopted, it is difficult to convince people that they constitute a social norm. However, other measures can be used to activate the idea of a social norm, without erroneously suggesting that the behavior promoted is common or widespread, such as the use of testimonials from other citizens whom the target audience of the communication are very likely to identify with (Elgaaied-Gambier *et al.* 2018). As suggested earlier, the call for opinion leaders to contribute can make it possible to bring about new routine behaviors in a progressive manner.

9.4. Conclusion

The present climate crisis leaves no room for further doubt of the need for public authorities to increase their involvement in the green transition. In this chapter, we have highlighted the contributions of marketing within the framework of applying environmental regulations.

According to Walsh (1991), the acceptability of market concepts in public services is limited, owing to the nature of the actions undertaken and the relationships between states and citizens. Marketing is certainly not the answer to everything, but it can be useful at different levels.

Based on the marketing mix and communications mix models, a regulatory mix or pro-environmental regulatory plan can be developed around five levers: coercion, incentives, taxation, awareness-raising and nudging (see Figure 9.1), with the challenge being to apportion each measure correctly and apply it at the opportune moment. Consideration focused on the regulatory mix makes it possible to reflect in terms of the market, by placing efficacy at the heart of pro-environmental measures.

Moreover, adopting a vision borrowed from megamarketing enables a holistic understanding of the different challenges for all actors, as environmental measures lead to resistance and disrupt power structures. Thinking in terms of alliances, supporters and opponents thus leads to a greater efficacy.

The marketing approach also invites the regulator to further question the importance of the psychological mechanisms of individuals, with the aim of developing relevant legitimization strategies. We also argue that segmentation and public action are not mutually exclusive, legislation does not exempt authorities and companies from communication, and convincing people is not limited to providing them with information.

That being said, such combinations of measures will only have a limited impact if they remained tied to a restricted geographical area. While local initiatives are to be encouraged, they are insufficient. As the climate crisis knows no borders, a transnational plan of action is essential. However, according to Nordhaus (2015), the efficacy of international climate agreements is running up against the phenomenon of "stowaways", in this case meaning countries that benefit from the reduction of emissions without contributing efforts to that end or bearing the costs of it. The author thus suggests establishing climate clubs, where non-compliance would lead to trade sanctions.

In any case, regarding simultaneous action involving several states, cultural specificities relevant to regulatory decisions must be taken into account. The country's political history, culturally-dependent psychological mechanisms and the prevailing environmental context are among the factors that can affect the nature of citizen responses to the different types of environmental regulations.

To conclude this chapter, we think it is crucial to indicate two priorities that will enable marketing researchers to support pro-environmental public policy decisions:

– *Explore new objects of research and new targets*: researchers in marketing are interested in brands and products. They study end and intermediate consumers. However, public policy measures could be considered as offers that should be accepted on a market. Marketing theories and methodologies could, for example, be used to isolate factors for the acceptance of an ecotax or coercive measures. Moreover, it would be interesting to study other types of decision-makers, such as elected representatives in local governments.

– *Adopt an institutionalist stance*: envisaging marketing as a legitimization practice enables a more effective understanding of the new objects of research mentioned in the previous point, in particular, by analyzing issues from the perspective of power and competing legitimacies. This perspective offers two advantages. First, environmental issues are not addressed from a reductive angle as a diversity of actors is taken into account to comprehend the success or failure of environmental policies. The second is that this definition of marketing is likely to make it more interesting in the eyes of decision-makers, thus making it more likely to be noticed.

9.5. References

Attari, S.Z., Schoen, M., Davidson, C.I., DeKay, M.L., Bruine de Bruin, W., Dawes, R., Small, M.J. (2009). Preferences for change: Do individuals prefer voluntary actions, soft regulations, or hard regulations to decrease fossil fuel consumption? *Ecological Economics*, 68(6), 1701–1710.

Ben Slimane, K. and Lecas, B. (2010). Le travail institutionnel : origines théoriques, défis et perspectives. *Management & Avenir*, 37(7), 53–69.

Bernard, Y., Bertrandias, L., Elgaaied, L. (2015). Shoppers' grocery choices in the presence of generalized eco-labelling. *International Journal of Retail & Distribution Management*, 43(4/5), 448–468.

Chen, X. (2011). Why do people misunderstand climate change? Heuristics, mental models and ontological assumptions. *Climatic Change*, 108(1), 31–46.

Elgaaied-Gambier, L., Monnot, E., Reniou, F. (2018). Using descriptive norm appeals effectively to promote green behaviour. *Journal of Business Research*, 82, 179–191.

François-Lecompte, A., Bertrandias, L., Bernard, Y. (2017). The environmental labelling rollout of consumer goods by public authorities: Analysis of and lessons learned from the French case. *Journal of Cleaner Production*, 161, 688–697.

Humphreys, A. (2010). Megamarketing: The creation of markets as a social process. *Journal of Marketing*, 74(2), 1–19.

Kaplan, A. (2007). Apports et limites du marketing dans l'administration publique. Master's report, ENA, Paris.

Kotler, P. (1986). Megamarketing. *Harvard Business Review*, 64(2), 117–125.

Moller, AC., Ryan, R.M., Deci, E.L. (2006). Self-determination theory and public policy: Improving the quality of consumer decisions without using coercion. *Journal of Public Policy & Marketing*, 25(1), 104–116.

Mullen, J.J. and Bowers, T.A. (1979). Government advertising: A runaway engine? *Journal of Advertising*, 8(1), 39–42.

Nordhaus, W. (2015). Climate clubs: Overcoming free-riding in international climate policy. *American Economic Review*, 105(4), 1339–1370.

Pelletier, L.G., Tuson, K.M., Green-Demers, I., Noels, K., Beaton, A.M. (1998). Why are you doing things for the environment? The motivation toward the environment scale (MTES). *Journal of Applied Social Psychology*, 28(5), 437–468.

Rhodes, E., Axsen, J., Jaccard, M. (2017). Exploring citizen support for different types of climate policy. *Ecological Economics*, 137, 56–69.

Schubert, C. (2017). Green nudges: Do they work? Are they ethical? *Ecological Economics*, 132, 329–342.

Suchman, M.C. (1995). Managing legitimacy: Strategic and institutional approaches. *Academy of Management Review*, 20(3), 571–610.

Thaler, R. and Sunstein, C. (2008). *Nudge: Improving Decisions About Health, Wealth, and Happiness*. Yale University Press, New Haven.

Tversky, A. and Kahneman, D. (1981). The framing of decisions and the psychology of choice. *Science*, 211, 453–58.

Vanclay, J.K., Shortiss, J., Aulsebrook, S., Gillespie, A.M., Howell, B.C., Johanni, R., Maher, M.J., Mitchell, K.M., Stewart, M.D., Yates, J. (2011). Customer response to carbon labelling of groceries. *Journal of Consumer Policy*, 34(1), 153–160.

Walsh, K. (1991). Citizens and consumers: Marketing and public sector management. *Public Money and Management*, 11(2), 9–15.

10

The Repairability of Household Appliances: A Selling Point for Utilitarian Products

Mickaël DUPRÉ[1], Patrick GABRIEL[2] and Gaëlle BOULBRY[3]

[1] *Governance and Insular Development, University of French Polynesia, France*

[2] *LEGO, IAE Brest, University of Western Brittany, Brest, France*

[3] *LP3C, University of South Brittany, Vannes, France*

10.1. Introduction

A French person throws away an average of 22 kilos of electrical and electronic waste every year (compared to a global average of 6 kilos) (Deloitte Développement Durable[1] 2016). It is therefore conceivable that the quantity of electrical devices discarded could be reduced. For companies, a product repairability policy requires them to meet at least two conditions: when developing the product, they must consider the (relative) ease of disassembling the product and replacing its parts; and ensure that parts are available for longer (seven to fifteen years for household appliances, for example) in a quick and inexpensive manner. Users, meanwhile, wish to reduce the amount of electrical waste: 77% of European citizens would rather repair their appliances than throw them away (Eurobarometer 2014).

1 French Environment and Energy Management Agency

Marketing for Sustainable Development,

Barriers to developing the repairability of products thus appear to be relatively small. However, it has not been easy to develop more active behaviors towards product repair: 44 % of appliances that end up discarded could have been repaired (BIO Intelligence Service S.A.S. 2012).

This chapter is focused on the effect of a lever that can be used by both manufacturers and users of household appliances to support product repairability: a "repairability" label. This aims to provide a better understanding of the effect of a "repairability" label on purchasing behavior in relation to household electrical products. This type of labeling is currently proposed by a number of industries in the form of a manufacturer's label (such as the "Repairable for 10 years" label used by Seb) or Ecolabels (such as the global Blue Angel label, which requires replacement parts to be provided for five years after the end of production).

The proposed study is focused on two essential dimensions of a manufacturer's commitment to the repairability of their products: the valence of the repairability guarantee and the effects on consumer purchasing decisions. More specifically, we will attempt to answer the following questions:

– How do consumers understand the concept of a commitment to repairability?

– What valence do they afford this commitment and the repairable product?

– Is repairability a selling point for household appliances?

– Is repairability a selling point for utilitarian products?

– Is repairability a selling point for people particularly conscious of environmental issues?

The answers to these questions are presented in two parts. The first allows us to clarify the concept of repairability and to summarize the favorable factors and the constraints related to this concept for manufacturers and consumers. The second presents the results of a study measuring the effects of a "repairability" label on the perceptions and behaviors of individuals purchasing household electrical products. In the conclusion, we will offer some recommendations that could enable such a label to enjoy broader acceptance among consumers.

10.2. Repairability: a complex concept

10.2.1. *Beneficial political incentives*

The repairability of products refers to the manufacturer giving users the possibility to repair their products or have them repaired after they have been used for a given period. This possibility of repair temporarily commits a company to its customers; this is also a sign of the company's societal responsibility. For Deloitte Développement Durable (2016), the repairability of products is part of a broader concept of "sustainable product", indicating a product's capacity to last. The term "sustainable" refers to *being strong, repairable, flexible and adaptable, compatible with other systems, easy to maintain, etc.* According to ADEME, such a policy would promote responsible consumption, one of the major obstacles to which is a general lack of durability (sturdiness, repairability, adaptability, etc.) that leads to a lower lifespan for some products.

A number of French and international bodies are campaigning for further information and communication on the lifespan of products. However, they are also questioning the conditions for effective communication and labeling. For Deloitte Développement Durable (2016) in particular, providing the consumer with more information could simultaneously increase trust, move towards more sustainable products and stimulate competition between manufacturers on the quality and longevity of their products. One such piece of information could also allow the consumer to better understand the usage costs of equipment; consumers do not always take the operating costs of products into account at the moment of purchase. Deloitte Développement Durable (2016) emphasizes in summary that any extension of the lifespan of objects requires greater trust, as well as an increased exchange of information between consumers and manufacturers. The working group of the French National Council on the Sustainability of Consumption and Products (*Conseil national de la consommation sur la durabilité des produits* – (Conseil national de la consommation 2015)) recommends, by including the particularities and challenges of activity sectors, enhancing consumer awareness, in particular by carrying out framed experiments on labeling information related to the lifespan of products on a voluntary basis. It also encourages communication on the repairability of products by indicating the availability of spare parts that are essential to the use of the good. Even Europe itself is questioning the conditions for labels indicating the lifespan of products and their repairability (see text Box 10.1). As a result, in 2017, the members of the European Parliament asked the European Commission to establish a "voluntary European label" that considers, among other things, the item's sustainability and the ability to replace its components in the interests of repairability.

In order to respond to the constraints highlighted by consumers[2] and denounced by a number of consumer associations (*UFC-Que Choisir, 60 millions de*

2 Generally through consumer associations, citizens have made public their reasons for not carrying out repairs: the difficulty of finding a repairer or the required parts; the impossibility of disassembling the device (pieces welded together, overly specific repair tools, etc.); the high cost of repairs, which is often too close to the price of a new item.

consommateurs) and pro-environmental organizations (*Les amis de la Terre*, for instance), the French government adopted legislative measures to promote responsible consumption: the Consumption Act requires the manufacturer to inform the seller of the date until which the spare parts needed to repair products introduced on the market since March 1, 2015 will be available. This information must be displayed at the point of sale. The manufacturer is also required to provide sellers and repairers, within a two-month timeframe, with the spare parts needed to repair products. The "Roadmap for the Circular Economy", published by the French government in April 2018[3], which was enacted by Law No. 2020-105 of February 10, 2020 on combating waste and on the circular economy[4], includes several measures on repairs. One of these is for it to be compulsory to display, from January 1, 2020, basic information on repairability for electrical and electronic items (Feuille de route 2018, p. 19). The aim is to strengthen the offering (availability of spare parts, training, etc.) and to encourage demand (mention of a repairability index, warranty, awareness-raising, etc.) to promote repairs. As a result, for the French government[5] and for ADEME, the recourse to the reuse and repair of products must become normal and attractive to the consumer (Feuille de route 2018, p. 19).

When, in 2012, I proposed, to the advisory body for the European institutions, the European Economic and Social Committee, of which I was then a member, that guidance be prepared on the issue of product sustainability, I had to be refused at first. The topic, which had not yet been considered by the European Union at all, appeared to be a minor and, above all, divisive one. It is true that its media coverage had been based around a conspiratorial idea of manufacturers sabotaging their own products themselves to increase sales. The guidance has been finally adopted and published in the *Official Journal of the European Union* on March 10, 2014. It allowed us to draw three main conclusions.

– The issue of product sustainability can lead to a very broad consensus. This was the case at the level of the European Economic and Social Committee, where the text was unanimously adopted by representatives of companies, trade unions, consumer associations and environmental NGOs, and some months later, in July 2017, at the European Parliament where all political parties, with the exception of extremist and populist parties, declared their support for decisive action; the dynamic in support of more sustainable products is now widely shared.

– After several years of stagnating initiatives, the European Commission is now taking on the subject again and the new circular economy action plan published on March 11, 2020

3 Available at: https://www.ecologique-solidaire.gouv.fr/sites/default/files/Feuille-de-route-Economie-circulaire-50-mesures-pour-economie-100-circulaire.pdf.

4 Available at: https://www.legifrance.gouv.fr/eli/loi/2020/2/10/TREP1902395L/jo/texte.

5 Available at: https://www.economie.gouv.fr/loi-consommation/mesure/consommation-responsable.

gives a strong importance to sustainability. The role and place of consumption practices within the circular economy is now fully acknowledged.

– The available means of communication must be used to the full here. Pioneering companies in product sustainability or repairability can thus win the loyalty of their customers by saving on their advertising costs. Moreover, the recent French law on combating waste and on the circular economy of February 10, 2020 has perfectly integrated the role of communications in sustainability policies. The Authority for the Professional Regulation of Advertising (*Autorité de régulation professionnelle de la publicité* – ARPP) also included the issue of sustainability in advertising associations in its recommendation of April 29, 2020. The fact that I had been missioned by the last two Green Transition Ministers, firstly, to work on sustainability and then to explore the role of advertising in the green transition clearly indicates the full recognition of the key role of communications, which, while it can make good practices more attractive, must also be carefully framed to avoid suspicions of greenwashing.

Box 10.1. *Planned obsolescence: A European perspective – interview with Thierry Libaert, Professor, Academic Contributor at the Earth & Life Institute (Catholic University of Leuven)*

10.2.2. *Environmental labeling: effects that are difficult to grasp*

Labeling on the lifespan and the length of availability of spare parts provides great hopes of changing consumption habits to make them more responsible ((Desbordes *et al.* 2014); recommendations of the French National Council on the Sustainability of Consumption and Products (Conseil national de la consommation 2015); ADEME study (2015); report by RDC Environnement (2017)).

Information directed at consumers is indeed a key factor in making the repairability of products a selling point and a purchasing criterion. This information must be both simple and credible to impart knowledge and build trust among buyers of repairable products. This is why the law on combating waste and on the circular economy (2020) will make it compulsory from 2021 to provide information on repairability in the form of an index on the model of the energy label.

However, a label is not a guarantee of either proper understanding or acceptance on the part of the consumer. A study carried out in October 2017 for the magazine *Que Choisir* among 2,206 of its readers on the influence of trade awards (such as "product of the year") on their consumption behaviors indicate that only a few labels are known or recognized (two out of eight were recognized either by name or by appearance by over half of the individuals interviewed); the proportion of people who trust the label[6] is even lower.

Labeling on product repairability increases the promise of a high value in use[7], and this promise affects the perception of the quality of the product. This halo effect, when the assessment of one attribute of an item influences the perceptions of its other attributes, has been observed in environmental advertising, for example (Lee *et al.* 2013). In this field, various experiments have shown that indicating that a product is "environmentally friendly" is enough to make consumers think that it tastes better (Fillion and Arazi 2002), is of a higher quality or performs better (Sörqvist *et al.* 2013) than an equivalent alternative. Environmental labels can influence the perception of a product, even features that have no direct connection to the label. For example, people think that fairtrade chocolate is of a higher quality (Schuldt *et al.* 2012) and tastes better (Lotz *et al.* 2013) than unlabeled alternatives, even though the fairtrade aspect has no bearing on the taste of the product.

10.2.3. *A limited selling point*

The main limitation of the repairability argument relates to the fact that the operating duration is not always the main element of the value in use that is sought by consumers. A study by Crédoc[8] (Tavoularis *et al.* 2015) highlights that the influence of this argument varies depending on the type of product. Quality is considered more desirable for household electrical products and food items. In the case of these categories, the willingness to pay more is higher for reliable, effective and safe products. Moreover, the same study points out that the search for quality in household appliances increases with consumer experience and age (and not according to generations).

Some categories of products are subject to a greater number of renewals, such as fashion products, rapidly-evolving technology products and, more recently, furniture products (Deloitte Développement Durable 2016). Consumers can also take into account that some items are now better-performing, more effective and more energy-efficient than previously; it is therefore more efficient and effective for them to buy these items regularly than to keep "used equipment" and repair it.

---

6 Available at: https://www.quechoisir.org/actualite-logos-de-consommateurs-infographie-quel-impact-sur-les-comportements-n49108/.

7 The value in use indicates the subjective value given to the advantages expected from the product use.

8 *Centre de recherche pour l'étude et l'observation des conditions de vie* – Research Center for the Study and Observation of Living Conditions.

The argument for repairability is thus subject to the established expectations of the user; repairability constitutes a product's positive attribute if the buyer wants a longer product lifespan or wishes to contribute to responsible consumption. In these two cases, it is necessary to look ahead (in terms of product usage or collective well-being) at the moment of purchase. Furthermore, a product's repairability is really only considered to the extent that the user is committed to repairing it; this requires the user to stop using the product for a time (to temporarily dismantle it) or even to participate in repairs. However, this commitment and cessation of use can be determined by psychological ownership that the user may have developed towards the object. For example, a study by Kreziak *et al.* (2016) identifies the conditions under which cell phone users will repair their products. The troubleshooting (and thus retention of the used phone) is more likely to be considered in cases where this will maintain all uses of the item. In economic terms, troubleshooting is considered when the residual value is utilitarian of nature. If the link with the brand (more so than with the phone itself) is more emotional, or when the possibility of reselling the used phone emerges (for example, in cases of financial takeovers by the operator), consumers tend to decide to separate from their former phone rather than have it repaired.

More generally, age seems to influence attitudes and behaviors in relation to the environment (Barber *et al.* 2014), but the results in the literature are contradictory. The effects of age are, in fact, indirect. While people trust more in labels with age (Teisl *et al.* 2008), the demand for green products is generally negatively correlated with age, with younger people having a greater awareness of environmental issues (Wessells *et al.* 1999). Some groups of consumers are thus not necessarily looking for products with a longer lifespan. Deloitte Développement Durable (2016) emphasizes that some categories of users deliberately choose products with short lifespans: for example, parents buying for their children whose tastes and uses will change rapidly, "nomadic" people who move often for personal or professional reasons, or even people who are conscious of fashion trends.

Beyond these limitations, the repairability of products can be hindered by contradictory beliefs. A study by Deloitte Développement Durable (2016) revealed that most consumers think that their new equipment is less resilient and that its lifespan is shorter. In addition, more than 90% of them distrust manufacturers, considering the lifespan of products (particularly technological ones) to be planned. Among some people, such a belief is so entrenched that communications on longer lifespans can be seen as lacking credibility.

In summary, while the repairability of products presents a number of favorable arguments at both the individual level (whether it is the company implementing it or the user) and the collective level (in terms of sustainable development), there are several constraints that can limit its application (see Table 10.1), to the extent that its implementation and acceptance by stakeholders cannot be taken for granted.

Arguments in favor of repairability	Arguments limiting repairability
It is a part of a "sustainable product" policy and thus encourages responsible consumption (ADEME 2016)	– Overload at the product design stage (having to consider the possibility of dismantling the product and replacing parts); make parts available over a long period (storing and managing parts) – The sustainability concerns required for the user to look ahead and commit to repair
Promises a long value in use and, through the halo effect, a high quality product	– Quality is not always what a consumer looks for in a product; this depends on the product category and the age of the buyer – The involvement degree of consumers towards the product has an influence: the residual value can count for more than the value in use in the decision to repair
Increases trust in the product, and in the company or brand	– The operating lifespan is not always what consumers are looking for – Lack of trust among consumers in the reliability of some products (new products versus older ones; planned obsolescence of some technological products)
Extends the length of the customer relationship and loyalty to the brand	– Quick replacement is desirable for some product categories (fashion items, products with rapid technological change, furniture) – Some consumer segments favor products with shorter lifespans

Table 10.1. *Elements supporting and limiting the repairability of products*

The repairability of products therefore appears to be a complex concept; it offers *a priori* a number of advantages for companies (trust, loyalty, perceived quality) and for society more broadly. However, it also has limitations and even risks of commercial failure. The dissemination of information between manufacturers and consumers on the possibilities for product repair seems to be a key element. A "repairability" label is a means of summarizing this information, and we must study its effects on consumers of household appliances more closely.

10.3. The effects of a "repairability" label on purchasing behaviors: mixed results

The study presented below was carried out in collaboration with the Seb group, a world leader in small electrical household appliances.

10.3.1. *The study: an experiment using fictitious e-commerce sites*

Our aim is to understand the effects of a "repairability" label on consumer purchasing behaviors. Specifically, we have tested the effect of mentioning a (fictitious) label guaranteeing the repairability of a product over a ten-year period (labelR) on online purchasing decisions. In order to do this, we conducted an experiment designed around two fictitious e-commerce sites, identical except for the presence of the labelR on some products. The simulation concerned three types of household appliances (coffee makers, irons, vacuum cleaners). The individuals interviewed therefore had to choose a coffee maker, an iron and a vacuum cleaner from among the range of products offered on the e-commerce site designed for the experiment. The products tested were representative of leading products on the French market. The methodological details of the experiment are presented in Box 10.2.

Figure 10.1. *Repairability label*

We will call a product presented without a label on the control site and with the labelR on the experiment site "product R".

The participants were contacted by email to participate in the experiment. By following a link sent to them, they were directed to a presentation page that explained the details of the experiment. They were then invited to simulate a purchase for an iron, a vacuum cleaner and a coffee maker.

The user experience was split into two distinct and successive tasks: a purchasing simulation and a questionnaire. The purchasing simulation was carried out through a site similar to a typical e-commerce site. By confirming their shopping basket, the participants were directed to a page with the questionnaire.

A first series of questions concerned the participants' beliefs about the responsibilities involved and valence of the labelR. A second series of questions was about awareness of environmental issues and made it possible to profile participants.

The experiment was carried out among a panel of 816 consumers, respecting the main quotas for representation of the French population (age, gender, profession and socioprofessional category, income, region).

Moreover, we conducted a study of 155 respondents in order to assess the utilitarian dimension of each of the three types of products (coffee maker, iron, vacuum cleaner).

Box 10.2. *The "repairability" label and the experiment on two fictitious e-commerce sites*

The results of the study allow us to understand how the labelR is viewed by consumers and its concrete impacts on purchasing decisions.

10.3.2. *Understanding labelR: a positive valence*

The labelR is very widely perceived as a commitment on the part of the manufacturer (80% of participants) to stock and provide spare parts for a period of ten years after the purchase of the product (66.4%). The participants also expect customer service (35.4%) and assistance in the repairs (52.2%). However, they are wary of the service guaranteed by the manufacturer. In fact, only a minority expect to be able to contact a repairer near them (23.3%) and to access inexpensive spare parts (28.4%).

The labeling also shows a positive valence in terms of the reliability of the product. It is perceived as reassuring and indicative of the significant reliability of the product (43.9%) and a reduced risk of breakage (38.1%). A minority of respondents see this as a sign that the product cannot be relied on and will break down (8.2%) or as a risk that the product will have to be repaired (15.4%).

In summary, the labelR seems to be understood by the majority of the individuals interviewed. Its effect goes beyond the repairability of the product itself to also encompass the reliability of the product. The labelR thus tends to reassure consumers.

However, the interpretation of the labelR remains imprecise, which is not surprising given its relative novelty. In particular, consumers implicitly associate a number of services with the concept of repairability. These services can be considered positively (assistance with repairs) or as more restrictive (the distance to get the object repaired, the cost of spare parts).

We also wish to know whether people who are more environmentally conscious are more influenced by the labelR than people with lower awareness.

The conclusions have been formulated on the basis of the responses of 252 participants that we have grouped into three similarly sized categories: low environmental awareness, moderate awareness and high awareness. The participants in each of these three categories selected the same average number of R products for their baskets (Chi2 [2, N = 252] = 0.476, $p = 0.79$). This result allows us to conclude that the labelR is not perceived as an ecolabel.

10.3.3. *The effects of the labelR on purchasing decisions: utilitarianism as a moderator*

For the range of products purchased, the statistical analyses indicate that mentioning the labelR has no effect on consumer choices. More specifically, the overall number of R products does not increase with the indication of a labelR ($p = 0.212$). The labeling will therefore not increase the purchase decisions in favor of labeled products. The number of participants that selected at least one R product among their three choices (vacuum cleaner, iron, coffee maker) also did not increase with the mention of the labelR.

We have been able to see that the three types of products tested are perceived differently in terms of their utilitarianism among the potential buyers (Chi2 [2.154] = 51.95, $p < 0.001$). The utilitarian dimension refers to the basic function of the household appliance. The participants consider the vacuum cleaner to have a significantly higher utilitarian function than the other two products; the vacuum cleaner is thus mainly considered a "cleaning device". The iron is, for its part, perceived as having a higher utilitarian dimension than the coffee maker. The perception of the latter thus tends to vary more than for the other two products; its purpose is to make coffee but it can have secondary functions such as, for example, to be a decorative object for the kitchen, or representing a certain lifestyle through its design.

This perception of the utilitarian dimension of a product has an impact on the effect of its repairability: the impact of the labelR is more important for the vacuum cleaner than for the other products. In concrete terms, the labelR increased the number of decisions to purchase an R vacuum significantly ($p = 0.053$). A trend was observed among the purchasing decisions for irons ($p = 0.098$) but no effect was seen in the case of coffee makers ($p = 0.263$).

We thus observed that the labelR has a significant effect for the most utilitarian product, a trending effect for the product with an intermediate utilitarian function and no effect for the least utilitarian item.

We have also compared, product by product, the average price of the product chosen on each e-commerce site (the control site versus the experiment site). The results conclude that there is a trending effect on the average price of the chosen vacuum and no effect for the other products. Indicating repairability on the label will therefore not encourage people to purchase more expensive products.

10.4. Conclusion

In order to promote a circular economy and contribute to improving the lifespan of products, the French government is relying heavily on making information available to buyers, especially simple and understandable information attached to products (including domestic appliances) in order to change purchasing and product usage behaviors.

The study carried out in this chapter testing the effect of mentioning a label (labelR) confirms a number of results provided earlier on the factors supporting and limiting the repairability of products; it also provides some clarification on the behavioral change hypothesis. We will focus on several points: the change in purchasing behavior; the perception of repairability; and the strategic challenge of repairability.

In terms of behavioral changes triggered by a "repairability" logo, the study shows that this is not automatic. This change seems of little importance and mainly occurs in the case of products that are seen as utilitarian (the vacuum cleaner in the study). This confirms the various observations of ADEME: the desired operating lifespan of products depends on their use, but, in the case of a coffee maker for instance, this use can be utilitarian (making coffee) as well as symbolic, if the coffee maker occupies a decorative position in the kitchen. The lifespan is thus doubtless less important.

While repairability, in the form of a logo affixed to products, does not convince everyone, it is recommended in marketing to address the part of the market *a priori* more open to this concept, before gradually extending it to other potential market segments. It may be surprising that the "repairability" label does not have a considerable influence on environmentally-conscious individuals. It would be interesting to explore whether a complementary effect exists between the labelR and an environmental label, such as the European Ecolabel or the *NF Environnement* label. Such a complementary effect could thus translate one form of synergy with a label into another: the environmental label would benefit the labelR by indicating respect for the environment, while the labelR could enhance the perceived quality of the household appliance, leading to increasing consumer trust in products. This type of effect has been seen, for example, with organic labels and the *Label rouge* for food products. In their study, Dufeu *et al.* (2014) show that association of a known label (the French AB label for organic products) with a little known label is greatly beneficial to the latter, leads to a reduction in consumer skepticism towards the information conveyed by the label and translates into a higher willingness to pay.

The knowledge and perception of a commitment to repairability on the part of the manufacturer remained relatively vague among the individuals interviewed. If they attribute this commitment to the manufacturer and perceive the product to be more robust, they are in the minority in considering the commitment to be easy to value for consumers (access to a repairer, low repair price). Companies should therefore carry out communication efforts to make the repairability concept and the labeling around it a reality, as well as more thorough work to build trust between companies and the public in relation to the responsible engagement of commercial organizations.

The results of the study also show that repairability is perceived as a matter for all stakeholders; it clearly requires action on the part of the manufacturer, which ensures the repair and the availability of parts, but it is also perceived as a commitment (constraint) on the part of the user, who is concerned by access to repairs (cost, time, distance). Repairability must therefore be developed in a spirit of "co-repairing". The recommendation for manufacturers is therefore to focus less on the product and concentrate more on the users and the relationship with them. This is why support services matter; whether they, for example, provide existing information (on the availability of parts, the repair process), or a recovery service (means of collecting appliances to be repaired – collection points, home service, etc.), a delivery service (once the product is repaired) or even a collaborative repair service (availability of a platform for "DIY-ers"), all of these services work together to make repair a reality for users and allow them to access it.

More generally, the challenge of a label on product repairability is perhaps not so much related to the growth of a market share as to the strategic potential of repairability.

One strategic challenge at the national level was identified by ADEME and the French government: guaranteeing the repairability of products could increase consumer trust in brands, as well as consumer loyalty (Kréziak *et al.* 2016), thereby strengthening the competitiveness of French and European companies in the faces of low-cost models based abroad. This differentiation opportunity has led Seb, a manufacturer of appliances, to develop a 10-year repairability guarantee under which it will ensure that all parts of the product can be disassembled and replaced, that the cost of repairing a part is reasonable and that replacement parts are available.

Repairability is also an argument for increasing consumer loyalty to the brand and it can enhance the image of a responsible company that is committed to combating product obsolescence and the creation of waste. This kind of responsible positioning has been taken by the outdoor clothing manufacturer Patagonia, emphasized by the slogan used in one of their advertising campaigns: "Don't buy this jacket". Through this slogan, the brand committed itself and the consumer to reimagining "a world where we take only what nature can replace"[9]. The company thus promotes the repair of its clothing, in particular by using a program called Worn Wear and highlighting the long and emotional relationships that consumers have with their favorite clothes, which they become attached to and can be in need of repair. More generally, Patagonia is developing the Common Thread initiative, which encourages stakeholders to find solutions, including repair, in view of reducing their environmental footprint.

The strategic challenge can be lastly understood at the level of each company, such as the possibility of strengthening links with stakeholders. The specialized retailers Darty and Leroy Merlin, for example, are developing collaborative spaces for creation/repair that are open to everyone.

Repairability seems therefore to be a fundamental concept that favors both responsible consumption and corporate social responsibility. However, this is a concept that, for the manufacturer, requires an examination of the use of its products by the consumer (fashionable, symbolic or utilitarian products) and of customer relationships. Repairability does not only concern the product; in its implementation, it requires a marketing program that is focused on developing links with users and on the co-creation of value.

---

9 Available at: https://eu.patagonia.com.

10.5. References

ADEME (2015). Réemploi, réparation et réutilisation. Synthesis [Online]. Available at: https://www.ademe.fr/sites/default/files/assets/documents/reemploi-reparation-reutilisation-2015.pdf.

Barber, N.A., Bishop, M., Gruen, P. (2014). Who pays more (or less) for pro-environmental consumer goods? Using the auction method to assess actual willingness-to-pay. *Journal of Environmental Psychology*, 40, 218–227.

BIO Intelligence Service S.A.S. (Mudgal, S.T., Tinetti, B., Faninger, T., Lockwood, S., Anderson, G.) (2012). Étude sur la durée de vie des équipements électriques et électroniques. Final report ADEME.

Conseil national de la consommation (2015). Avis du Conseil national de la consommation sur la durabilité des produits. No. NOR: EINC1511471V, May [Online]. Available at: https://www.economie.gouv.fr/files/files/directions_services/cnc/avis/2015/Avis_sur_la_durabilite_des_produits.pdf.

Deloitte Développement Durable (Monier, V., Deprouw, A., Jover, M., Chouvenc, S.) (2016). Rapport annuel du registre des déchets d'équipements électriques et électroniques. Annual report, ADEME, 2015.

Desbordes, V., Font, E., Guillaume, E., Papineau, P., Perdereau, J., Ravet, A., Sinrat A., Ravet, C. (2014). Assistance dans l'élaboration d'une stratégie nationale relative à la durée de vie des produits, du réemploi et la réparabilité qui contribuent à des économies de matières premières, de CO_2 et de déchet. Study report 011433, document DE/3, Laboratoire national de métrologie et d'essais, Paris [Online]. Available at: https://www.lne.fr/publications/guides-documents-techniques/rapport-etude-duree-vie-produits-lne-juin-2014.pdf.

Dufeu, I., Ferrandi, J.M., Gabriel, P., Le Gall-Ely, M. (2014). Multi-labellisation socio-environnementale et consentement à payer du consommateur. *Recherche et applications marketing*, 29(3), 34–55.

Eurobarometer (2014). Attitudes of Europeans towards waste management and resource efficiency. Flash Eurobarometer 388. Report, European Commission, Brussels.

Fillion, L. and Arazi, S. (2002). Does organic food taste better? A claim substantiation approach. *Nutrition and Food Science*, 32, 153–157.

Kreziak, D., Prim-Allaz, I., Robinot, E., Durif, F. (2016). Obsolescence perçue, décision de renouveler et destinée des produits : le cas du téléphone portable. *Décisions Marketing*, 81, 41–59.

La Fabrique Écologique (2016). Comment agir vraiment contre l'obsolescence programmée ? Aller vers une consommation plus soutenable. Summary, La Fabrique Écologique, December.

Lee, W.C.J., Shimizu, M., Kniffin, K.M., Wansink, B. (2013). You taste what you see: Do organic labels bias taste perceptions? *Food Quality and Preference*, 29(1), 33–39.

Lotz, S., Christand, F., Fetchenhauer, D. (2013). What is fair is good: Evidence of consumers' taste for fairness. *Food Quality and Preference*, 30(2), 139–144.

RDC Environnement (2017). L'obsolescence programmée : politiques et mesures belges de protection du consommateur. Final report, RDC Environnement, May [Online]. Available at: www.marghem.be/wp-content/uploads/Obsolescence-programm%C3%A9erapport-final_RDC-Envrionment_V2_Rapport.pdf.

Schuldt, J.P., Muller, D., Schwarz, N. (2012). The "Fair Trade" effect: Health halos from social ethics claims. *Social Psychological and Personality Science*, 3(5), 581–589.

Sörqvist, P., Hedblom, D., Holmgren, M., Haga, A., Langeborg, L., Nöstl, A. (2013). Who needs cream and sugar when there is eco-labeling? Taste and willingness to pay for "eco-friendly" coffee. *PLoS ONE*, 8(12), 807–819.

Tavoularis, G., Hébel, P., Billmann, M., Lelarge, C. (2015). Comment a évolué sur les deux dernières décennies la relation à la qualité pour les consommateurs français ? *Cahier de recherche*, 327.

Teisl, M.F., Rubin, J., Noblet, C. (2008). Non-dirty dancing? Interactions between eco-labels and consumers. *Journal of Economic Psychology*, 29, 140–159.

Wessells, C.R., Johnston, R.J., Donath, H. (1999). Assessing consumer preferences for ecolabeled seafood: The influence of species, certifier, and household attributes. *American Journal of Agricultural Economics*, 81, 1084–1089.

11

The Role of the Fairtrade Label in the Spread of Sustainable Production and Responsible Consumption in West Africa: The Case of Côte d'Ivoire[1]

Mantiaba COULIBALY-BALLET
GRM, Côte d'Azur University, IAE Nice, France

11.1. Introduction

The year 2020 was marked by a health crisis caused by the Covid-19 virus and every country in the world has been affected. The impact of this virus is beginning to be felt, in particular among producers throughout West Africa, as well as in Central/South America and the Caribbean. One of the main concerns relates to the continued drop in cocoa prices (down approximately 25% since 2016). In addition, the transportation and export of this raw material are increasingly difficult owing to border closures (Fairtrade 2020). Alongside this health crisis are the consequences of climate change: increasing drought, rising sea levels, the acidification of the oceans, coastal erosion, etc. (Clairouin *et al.* 2014). Moreover, cocoa cultivation in Africa is often responsible for the destruction of native forests, owing to fires caused

1 The author wishes to warmly thank Sihem Dekhili for her comments and suggestions, as well as for the numerous re-reads of this chapter to enhance and improve its content.

by farmers. The consequences of climate change remind all countries of the importance of sustainable development.

In addition to these challenges, Africa is facing other issues, such as the significant growth of its rural populations (Vimard and Fassassi 2011) and difficulties at the production level for some products, such as coffee and cocoa. More specifically, the issue of remuneration of producers in West Africa often arises, especially as this region accounts for 70% of global cocoa production. According to Sébastien Maes of Oxfam-Magasins du Monde:

> West Africa is the biggest producer of chocolate and Europe is the biggest consumer… [However] we estimate that the producers earn only a third of what they should to have a so-called minimum necessary income, to have access to housing, education for their children and healthcare, as well as savings in case of hardship. (Radio Télévision belge francophone 2019)

Moreover, children can be made to work in cocoa fields. According to Maes, "harvesting periods, which more or less coincide with the beginning of the school year, that is, September–October, are the moments when there is a significant need for workers. It is then children who, instead of going to school, find themselves on these plantations".

In this context, how can sustainable food be provided for a fast-growing population while confronting issues in production, remuneration and poverty? Can fairtrade be a solution to providing a more equitable remuneration for cocoa producers?

Fairtrade is governed by standards (social, environmental and economic) that are often in line with the principles of sustainable development at both the environmental and social levels (Makita 2016). These standards affect all of the market actors involved in the fairtrade process, from the producer to the company that sells the fairtrade product.

Many producer cooperatives in the Pacific, Caribbean, India and Africa have now been certified with fairtrade labels, such as the Max Havelaar Fairtrade label, the oldest and most established label, which was launched in 1988. However, we have seen an increase in the number of labels in recent decades (Fair Trade USA, Biopartenaire, *Association pour le tourisme équitable et solidaire*, the Fair for Life program, *Symbole des petits producteurs*, Naturland Fair), leading to a lack of visibility of fairtrade products and to consumers questioning the guarantee of the label (Guide International des Labels du Commerce Équitable, GILCE 2020).

In Africa, Côte d'Ivoire is one of the countries that has made the most effort in terms of sustainability and the adoption of fairtrade certifications, especially in the cocoa sector. Almost 200 cooperatives, bringing together more than 120,000 producers, currently have fairtrade certifications in Côte d'Ivoire, compared to only one cooperative in 2004 (Le Point 2018).

Even though sales of fairtrade cocoa only represent approximately 7.5% of the national production of Côte d'Ivoire, they have doubled within a year, going from 80,000 tons in 2016 to 150,000 in 2017. Cocoa appears in most voluntary sustainability programs, such as "Cocoa Life" (launched in 2012, dedicated to supplying Mondelēz Cocoa) or "Cocoa Plan" (launched in 2012, dedicated to supplying Nestlé). In 2016, Côte d'Ivoire had 70 cooperatives and 40,000 cocoa farmers involved in the fairtrade program (GILCE 2020).

In this context, we are trying to find out how the application of fairtrade principles by Fairtrade certified producer organizations can contribute to sustainable consumption and responsible consumption.

This chapter analyzes the impacts of the application of fairtrade principles by producer organizations in Côte d'Ivoire. In the first section, we will identify the basic principles of the Fairtrade label and its potential implications for sustainable development and responsible consumption. In the second section, we will explore the application and implications of these principles in the context of Côte d'Ivoire.

11.2. The Fairtrade label: towards sustainable production and responsible consumption

In this section, we will first analyze the position of the Fairtrade label in relation to other fairtrade labels by highlighting the sustainability of production (section 11.2.1). We will then study the advantages of the Fairtrade label in terms of responsible consumption (section 11.2.2).

11.2.1. *The position of the Fairtrade label: the quest for sustainable production*

This section will be organized around two key points: 1) presenting the specificities of the Fairtrade label in a context of multiplicity of cues related to fairtrade; and 2) exploring the role of this label in encouraging sustainability.

11.2.1.1. *The Fairtrade label, one cue among many certifications…*

Since the late 1980s, fairtrade actors have used the label to distribute fairtrade products, including in non-specialized stores. It is thanks to this specific recognition system that fairtrade products are sold in all kinds of stores, such as large and medium-sized supermarkets where the vast majority of food purchases take place. There are several fairtrade labels and they are mainly private labels. They are defined by their visual identities (logos) and refer to specific standards relating to fairtrade principles (GILCE 2020).

Max Havelaar, the first fairtrade label, appeared in 1988 in the Netherlands. It was then followed by national labeling initiatives in various countries. From 1997, these initiatives were grouped together in a single entity: Fairtrade International.

Brands using the Fairtrade label are not necessarily devoted entirely to this type of commerce; it can be used by conventional companies whose activities are only partially dedicated to fairtrade. Social economy organisations, strongly involved in fairtrade, have also chosen to use labels to improve the credibility of their offerings. These companies aim allow their partner producers to benefit from the advantages of fairtrade.

Since the early 2000s, new labels have been developed with different views on the implementation of fairtrade. As a result, in 2013, a guarantee system was created by organizations specializing in fairtrade and that are members of the World Fair Trade Organization (WFTO). This guarantee system focuses on specific criteria concerning audits and certification costs that must remain affordable for small producer organizations (GILCE 2020).

In Table 11.1, we summarize the main fairtrade labels by specifying their creation dates, objectives and certification bodies.

We put an emphasis on the Fairtrade label as it is the oldest and the best known label throughout the world. This choice is also justified by the fact that, despite the development of new labels, Fairtrade remains the main actor in the sector and acts as a point of reference (several fairtrade labels mention the minimum prices of Fairtrade).

Labels	Creation date	Key objectives	Certification bodies
Fairtrade International **(Max Havelaar, launched on the initiative of the Dutch development agency, Solidaridad)**	1988	Strengthen the position of small producers: a bigger majority of members of cooperatives should be made up of small family farms (new standards for 2019 for small producer organizations). Follow the United Nations Sustainable Development Goals and the aims of the 2015 Paris climate agreement (COP21).	FLO-CERT accredited by Fairtrade International to implement the certification of Fairtrade specifications. Accredited ISO 17065
World Fair Trade Organization (WFTO) Original name: International Federation for Alternative Trade (IFAT)	1989	Facilitate the sharing of information and ideas related to fairtrade. The founding members are dedicated to supporting marginalized artisans in the South by helping them to access the north's markets. WFTO is now dedicated to supporting organizations committed to fairtrade values.	WFTO recognizes the following certifications as a guarantee that the suppliers comply with WFTO specifications: 1) SPP (*Symbole de petits producteurs* – small producer symbol) (FUNDEPPO); 2) Fairtrade International/FLO-Cert; 3) ECOCERT/IMO (Institute for Marketecology) "Fair for Life"; 4) Naturland FairTrade standards
Fair Trade USA (FTUSA) Or TransFair USA, was a member of Fairtrade International (Fairtrade labelling Organizations International: FLO). FTUSA is a non-profit organization registered in the United States	1998	Stimulate sustainable development and the emancipation of communities by encouraging a fairer model of global trade that benefits farmers, workers, consumers, industry and the planet.	The certifier Global Services SCS (Scientific Certification Systems) carries out audits for FairTrade USA. The services of Control Union are also used.

Labels	Creation date	Key objectives	Certification bodies
Biopartenaire: a French association	2002	Federate and support small and medium-sized companies which are certified *Biopartenaire*; link operators with sectors upstream and downstream; foster closer links between companies and professional networks of organic agriculture and fairtrade; and take part in work to raise awareness of these topics.	Since 2018, the *Biopartenaire* label has been based on two different standards: – the Fair for Life standards managed and monitored by Ecocert, specially adapted to international partnerships; – the FiABLE standards that make it possible for monitoring to be carried out by a certification body (external audit) and by *Biopartenaire* (internal audit).
***Association pour le tourisme équitable et solidaire* (ATES): non-profit association**	2006	Act for equitable, and solidarity tourism.	Evaluation of the label is carried out by external certification bodies rather than by peers.
The Fair for Life program: the product of a partnership between certifying bodies - IMO (Institute for Marketecology) and the Bio-Foundation	2006	Adopt a means of recognizing other certification systems that share the same principles and that have comparable monitoring measures. Recognition of ingredients certified by the following labels is possible: Fairtrade International, FairWild Foundation, SPP, FTUSA and Naturland Fair.	In 2016-2017, the IMO Fair for Life and Ecocert Fair Trade standards merged. They became Fair for Life for fairtrade and responsible supply chains, which is managed by Ecocert SA.
***Symbole des petits producteurs* (SPP) SPP Global is a non-profit organization**	2006	Only work with small producer organizations and exclude private plantations and non-organized small producers.	10 certification bodies are accredited by SPP Global: Certimez, Biolatina, Mayacert, BCS-Öko, Imocert, Biotrópico, Control Union Peru, Ceres, Quality Certification Services and TERO.

Labels	Creation date	Key objectives	Certification bodies
Naturland Fair	2010	Combine organic farming and fairtrade under one logo, applied to products from both the north and the south. This label certifies products from companies that have already been certified by Naturland standards on the social responsibility requirements of companies and organic production.	Naturland is a certification body accredited by the International Organic Accreditation Service (IOAS) under standard ISO/IEC 17065.

Table 11.1. *Overview of fairtrade labels (source: GILCE (2020))*

11.2.1.2. *The Fairtrade label: specificities encouraging sustainability*

The Fairtrade label has its roots in the “Fairtrade system”, which is based on three types of organizations: FLO eV (Fairtrade Labelling Organizations International, “eingetragener Verein”, a non-profit body), the limited-liability company FLO-Cert and national labelling initiatives (Balineau and Dufeu 2012).

FLO eV is the owner of the Fairtrade brand whose logo is affixed on products that respect its criteria. Its main role is to develop the different specifications that characterize this brand.

There are two main types of standards related to the Fairtrade label: standards for small producers and standards for traders. The former specify the criteria that the producer must satisfy to have access to the given trade conditions. These standards can be different for independent producers and producers organized into cooperatives. Overall, these requirements must make it possible to prove to consumers that the extra price they will pay will offer producers possibilities for social, economic and environmental development (Balineau and Dufeu 2012).

Standards for traders are applied, in particular, to actors who buy and sell Fairtrade products. They define the rules on the composition of the product, traceability and use of the brand. These standards also concern the requirements for voluntary best practices that ensure that Fairtrade products are traded in a fair and sustainable manner, and contribute to the growing autonomy of producers (Fairtrade 2020). In order to have the right to display the Fairtrade logo on their products, traders must pay the producers a "fair price" and pay their organization a "development premium". The fair price and the development premium are fixed by FLO eV for every world region and are revised annually. Traders in the Fairtrade system are also called on to commit to long-term relationships with producers and to provide them with pre-financing if required (Balineau and Dufeu 2012).

FLO-Cert certification inspectors check on producer organizations and traders every year to ensure the proper application of standards. In producer organizations, the inspectors are specifically examining whether accounts and minutes of general assemblies are being kept properly, whether bureau members are being rotated and whether the use of pesticides is reasonable.

The initial fair trader (the cooperative or import/export company) is also subject to close scrutiny to ensure that they are paying a fair price and development premium to producers. Downstream of the chain (in coffee-roasting companies in particular), the monitoring is sometimes considered less strict as FLO-Cert is largely limited to checking working conditions.

The concessionary of the Fairtrade brand (the company that puts the logo on products) is the last actor to be monitored (for example, Malongo for coffee). Only the end retailer (generally specialized stores) is not subject to any conditions. For all of the other actors in the chain, the possibility of selling fairtrade products depends on the reports of FLO-Cert inspectors (Balineau and Dufeu 2012).

The Fairtrade label has three main aspects that refer to the principles of sustainable development: social, environmental and economic (Coulibaly-Ballet 2019).

The social aspect aims to strengthen the power of small producers and workers through organization into democratic communities (within cooperatives), incentives for union organization (on plantations), regulated working conditions and the prohibition of child labor and discrimination (Max Havelaar 2019). The social issue emphasizes equal rights and the management of vulnerabilities, in particular. On the one hand, the principle of equal rights is intended to promote equality for all without discrimination. On the other hand, the principle of vulnerability management encourages support for a decent standard of living by taking into account the capacities of individuals and households to reduce their vulnerability (Sirdey 2014).

The environmental aspect, for its part, concerns environmental protection. The main actions in this area relate to the development of organic farming, the protection of natural resources, the prohibition of dangerous pesticides and the rejection of genetically modified seeds.

Lastly, the economic aspect highlights the requirements that are imposed on traders and producers at various levels: the payment of the minimum price and Fairtrade premium; the regulation of financial flows and goods; trade relations; pre-financing; and use of the Fairtrade label (Max Havelaar 2019).

The Fairtrade minimum price is extremely useful as it protects producers from the volatility of the global markets by establishing a minimum amount that must be paid to them (FLOCERT 2019). It also makes it possible to cover sustainable production costs and ensure that the basic needs of the producer family are met. However, in recent years, the minimum price has been challenged; the fact that a cooperative is included in a list of Fairtrade-certified operators does not mean that it will sell its produce in line with Fairtrade terms. This all depends on the cooperative's orders and the willingness of its buyers to appreciate the quality that the label reflects, by paying national associations for a Fairtrade license (Max Havelaar France, for example) (Ruf *et al.* 2019). In regards to the Fairtrade primium, this allows farmers to receive a set amount for all products sold under Fairtrade terms and to invest in areas considered important by producers; this can include local infrastructure, healthcare, schools, social projects to make their community more sustainable and improvements to their companies by, for example, buying drying equipment (Dekhili *et al.* 2017; FLOCERT 2019).

11.2.2. *Fairtrade and responsible consumption: a label in search of legitimacy among consumers*

This section will be structured around the following points: 1) an overview of the advantages of the Fairtrade label for the consumer and its limitations, as well as the sources of doubt; and 2) a proposal for areas to improve the legitimacy and effectiveness of the label.

11.2.2.1. *The Fairtrade label: advantages tarnished by consumer skepticism*

The Fairtrade label plays an important role as it takes into account both the quality of the product delivered and the well-being of the consumer in society (Coulibaly-Ballet 2019). It is often perceived as a social label because it "provides the consumer with a guarantee of respect for the social environment, i.e. the conditions of production and trade with producers" (Michrafy 2013, p. 30). The Fairtrade label is also considered a credible means of informing consumers of the invisible aspects of a product's quality because it comes from an independent third party. The company, the certifying body and the producer thus collectively commit to providing the consumer with a product that fulfils the specific criteria of the label.

Despite the cost of fairtrade products (on average 10–12% higher than their conventional equivalents), some consumers see these prices as fair as these products bring individual benefits, such as a higher quality, good taste and being more healthy. Other altruistic benefits also explain the acceptance of the price of fairtrade products. Fairtrade is associated with environmental advantages, such as the preservation of nature, and other social benefits, including improved working conditions for producers and protection for children (Dekhili *et al.* 2017).

The health crisis caused by the Covid-19 virus seems to have increased consumer interest in fairtrade products. According to a study carried out in France by the national movement of consumers supporting fairer trade and a fair and sustainable consumption ("FAIR(e)), during the first period of lockdown (between March and May 2020), French people showed a preference for healthy products with a better taste, irrespective of price (Dubreuil 2020).

However, despite the advantages of the Fairtrade label, consumers have a number of doubts about the fairtrade offer, which may explain the low conversion of consumer discourse in favor of fairtrade into actual purchases (De Pelsmacker and Janssens 2007). This confirmed attitude–behavior gap is largely explained by two kinds of uncertainty experienced by consumers: uncertainty around compliance and around the effectiveness of the standards. It is established that, fairtrade products are distinguished by features that are intangible (the conditions of production and trade, for example). The consumer is therefore unable to verify these features through a search for information before purchasing the products, nor during the consumption experience, or even after the products' usage (Castaldo *et al.* 2009).

Furthermore, consumers see some inequalities in fairtrade in the sense that some intermediary actors can profit at the expense of producers. They also worry about the existence of a compromise between certifiers and supermarkets (Dekhili *et al.* 2017). More broadly, the multiplicity of labels and institutions that proclaim their involvement in fairtrade tends to create confusion among consumers (Diaz-Pedregal 2007) and a lack of trust in fairtrade offerings.

11.2.2.2. *Some areas for increasing the legitimacy of the Fairtrade label and reducing consumer doubts*

Despite the guarantees offered by the Fairtrade label, a loss of legitimacy of this cue has been observed among consumers.

In order to maintain the legitimacy of fairtrade on the market, it is in the best interests of companies to create a sharing space for marketing and sustainable development by taking into account the specificities of each field. In marketing, it is short-term needs (expectations for speed and profitability) and a search for competitiveness (differentiating from and outperforming the competition) that dominate. Meanwhile, in sustainable development, it is progressiveness, prudence (not cutting corners, building solid bases, moderating requirements of margins and turnaround times) and a search for synergy (cooperations, alliances and partnerships) that are important. There is therefore a need for an area of overlap between marketing and sustainable development that takes three elements into account: 1) the importance granted to the concept of need (ability to identify and respond to them); 2) the need to plan (in connection with future generations and future markets); and 3) the search for an optimum balance between stakeholders (dialog, negotiation and mutual benefits) (Louppe 2006).

This area of overlap between marketing and sustainable development can be applied to the case of the fairtrade market and the Fairtrade label in the following way:

– *the importance given to the concept of need*: this involves carrying out studies to identify and respond to the real needs of consumers. This is the case for the recent study conducted by FAIR(e), with the support of a responsible consumption app (MyLabel) and the participation of Max Havelaar France. This shows a change in the consumption practices of French people. These are consumers who are increasingly looking for healthy and high quality products. This is thus a question of improving the perceived quality of fairtrade products and continually assuring the consumer of the merits of fairtrade, in particular in comparison to other responsible products, such as organic products. Improving the quality of fairtrade products involves increasing the environmental benefits of fairtrade by managing natural resources on Fairtrade-certified plantations more effectively, ensuring that producers are able to better adapt to climate change, banning GMOs and more. Compliance with the social and economic requirements of fairtrade should also be ensured (Max Havelaar France 2020);

– *the need to plan*: this element can involve the development of new markets, such as North-North or local fairtrade markets, as has been the case in France for five years. Local fairtrade valorizes regional products and local producers to allow them to make a living from their produce. Local trade also provides support for family agriculture, which often offers high quality products. Planning can occur in this context in the form of preserving and protecting natural resources for future generations. This is the case with reforestation activity in cocoa fields in Africa;

– *the search for an optimum balance between stakeholders*: consultations can be held between actors (producers, consumers, civil society actors and even retailers) to present the benefits of fairtrade and the mutual gains of each stakeholder. This is the case for exchanges that have taken place each year during the Fair Trade Fortnight in France. In addition, support for different actors, such as communities and schools, can be organized. The association FAIR(e) is thus strengthening its programs to raise awareness among actors by communicating on the accessibility of fairtrade products for consumers. The idea is to show the consumer what can be gained by consuming fairtrade products (products that are high quality, healthy, environmentally friendly, etc.). Such communication also constitutes a means of addressing the high price of these products, which could be a consumption barrier (Dubreuil 2020).

Moreover, a marketing mix approach can help to strengthen the legitimacy of labels and thereby encourage responsible consumption. Table 11.2 shows the possible ways of integrating the objectives and social and environmental impacts at the level of each component of the marketing mix (Louppe 2006).

Components of the marketing mix	Integration of objectives and social and environmental impacts
Product	Select suppliers and supply chains with CSR performance; adopt transformation and production processes which respect both people and the environment; improve the recyclability of products and equipment at the post-consumer phase.
Price	Justify price increases by incorporating social and environmental measures into cost structures (sharing the effort with the consumer); provide pricing incentives to consumers to encourage them to adopt eco-freindly behaviors (for example, purchase rebates for public transport users); enable the consumers to benefit (lower prices) from the savings achieved thanks to their responsible behavior.
Place	Carry out recruitment and training to improve reception functions; ensure that sales stores and reception spaces comply with ISO 14000 or HEQ (High Environmental Quality) standards; facilitate the recovery of used material and the return of containers for customers; choose engaged retailers.
Promotion	Integrate "soft prospecting" methods (less aggressive, energy-saving, etc.); provide guidance to consumers for a more responsible use of equipments and resources available to them (reduce wasting of water and electricity, respect of places and safety measures, etc.); communicate regularly on all improvements made through the involvment of all actors, and on efforts to resolve ongoing problems.

Table 11.2. *Integration of the objectives and social and environmental impacts into marketing programs (source: Louppe (2006))*

11.3. The application of the Fairtrade label by producer organizations in Côte d'Ivoire: challenges and implications

11.3.1. *Case study*

In this study, we aim to understand the contribution of fairtrade principles to sustainable production and responsible consumption by considering Fairtrade-certified producer organizations.

We have focused on the West African branch of Fairtrade Africa, and more specifically on the case of Côte d'Ivoire. The main reason for this choice is the significant development of the Fairtrade label in this country in recent years. Côte d'Ivoire has placed great importance on fairtrade for a number of years, especially for cocoa, coffee and cashew nut products. In addition, Côte d'Ivoire is taking an increasing interest in environmental protection, a sensitive issue for a country that has lost almost 90% of its forests in the space of 50 years, due to the combined effects of an increase in population (from 3 million people to 25 million) and agricultural development (Le Point 2018).

We have adopted a qualitative approach to collect information on producer organizations that use the Fairtrade label. These organizations work with Fairtrade Africa, which is a member of Fairtrade International.

Fairtrade Africa was created in 2005 and is split into four African regions: East Africa, Southern Africa, West Africa and North Africa. It aims to provide farmers with a better position in the market and support them in establishing a sustainable development approach in Africa. Fairtrade Africa is made up of over 300 African producer organizations that fulfil international fairtrade criteria. It involves small producer organizations and plantations that mainly produce coffee, cocoa, tea, cut flowers, bananas, pineapples, mangos, shea butter, honey, vegetables and dried fruits (Representative of Fairtrade Africa 2019).

We have been able to establish contact with four organizations: a banana plantation, an organization that produces and exports coffee and cocoa, an organization producing cocoa and coffee, and an organization producing honey. We then carried out face-to-face interviews with these organizations and their partners between June 2018 and April 2019, both during our travels to Côte d'Ivoire and remotely. A total of 19 semi-directed interviews were carried out, with each lasting between 30 minutes and 1 hour.

The respondents included: managing directors at Fairtrade Africa; an assistant to the head of the Fairtrade premium management committee in charge of quality, hygiene and the environment; a head of human resources; a head of the management committee of a banana cooperative; a head of sustainable development and supply in charge of the management of the coffee and cocoa producer organizations; an export assistant; a producer member of a cocoa-coffee cooperative; an assistant in charge of relationships between producers and customers; and a beekeeper (member of a honey producer organization).

During these interviews, we collected information on the producer organizations and their partners, and on their products. We have particularly emphasized the implementation of the economic, social and environmental principles of fairtrade, and its benefits for producers.

11.3.2. *Accompanying actions for producers: a source of sustainability and responsible consumption*

The four organizations studied support producers as part of the Fairtrade certification procedure. Each organization is composed of producers who are organized into small groups.

In terms of production, support actions cover three levels: social, economic and environmental. Moreover, other measures are aimed at developing responsible consumption in Côte d'Ivoire.

11.3.2.1. *Accompanying actions at the social level*

Support is provided to producers who are members of a cooperative, as is the case for cooperatives of banana plantations: "We have insurance that covers the family, female empowerment (making women independent), accommodation, free water" (assistant to the head of the Fairtrade premium management committee, banana plantation). Support at the social level also involves the integration of women into the workforce: "We have an association of female producers; they put the hives in place and the men collect" (assistant in charge of relationships between producers and customers, honey producer organization). Another aspect concerns the rights of children. Even if awareness of this issue emerged well before the certification project, actions have since been strengthened, particularly in terms of prohibiting child labor: "Children do not work in farm fields; there can only be socialization work (like learning, or training children on growing products, basics, seasons, etc.) that children do" (producer, member of a cocoa-coffee cooperative).

11.3.2.2. *Accompanying actions at the economic level*

Producer organizations can obtain benefits, such as access to the certification and receipt of a premium. On the one hand, Fairtrade International enables greater access to certification fees: "The costs are acceptable in comparison to other labels; there is trust there" (head of sustainable development and supply, coffee-cocoa producer and export organization). On the other hand, it makes it possible to increase the premium received by producers: "Fairtrade is well regulated; the premium will be going up in comparison to other certifications, such as Rainforest, Utz, USDA organic" (producer member of cocoa-coffee cooperative). This benefit enables certified producer organizations and cooperatives to invest in production equipment and in social projects (schools, free clinics, etc.). It was reassessed in October 2019. As a result, an amount of $240 per ton of produce is paid to certified cooperatives and each cooperative decided how to invest that money at their annual general assembly. In addition to this premium, the Ivorian government was committed to implementing its own mandatory income premium in October 2020. This new financial advantage, which amounts to $400 per ton of cocoa, will ensure greater stability and security for producers and families (Max Havelaar France 2020).

11.3.2.3. *Accompanying actions at the environmental level*

Producer organizations support farmers to provide them with training in environmental protection and to increase the traceability of products.

Broadly speaking, the traceability of products concerns knowledge of the journey of the food product throughout the production and distribution chain (from the field where it was grown to the store where it is sold, via the place where it was processed or stored). Thus, risk control can be carried out in case of a health incident, as the quality of the products can be assured and the origin easily identified. In the context of the Fairtrade label, traceability consists of isolating fairtrade ingredients from other ingredients at each stage of the manufacturing, processing and distribution process. This involves storing and processing these ingredients separately.

Producer organizations also support farmers in carrying out their identification. The fairtrade producers are directed to authorized growing areas to preserve protected forests: "The identification of producers is done so that they avoid listed forests" (head of sustainable development and supply, coffee-cocoa producer and export organization).

Organizations also provide training for producers to help them to address issues relating to the contamination of fields and climatic events. The training also includes environmental protection techniques: "training in honey production, training in fairtrade and an envirnmental education. All of this avoids contamination" (assistant in charge of relationships between producers and customers, honey producer organization).

We note that plantations can become contaminated by some insects or by the use of harmful products. The training thus leads to awareness-raising programs: "We raise workers' awareness of contamination, carbon footprints, environmental issues" (assistant to the head of the Fairtrade premium management committee, banana plantation); "Awareness-raising tours are organized to prevent contamination; herbicides are banned. Within a village, almost everyone is a member of the cooperative, which avoids contamination" (producer, member of a cocoa-coffee cooperative).

There is also an emphasis on raising awareness among producers of the harmful consequences of flooding and drought, which allows the prevention of environmental risks: "We are in training. In 2014, we had floods and again four years later (2018). So there is an interest in sustainable development; it's become an obligation" (assistant to the head of the Fairtraide premium management committee, banana plantation); "Awareness-raising is carried out to inform producers how to manage things properly during droughts; we have natural climatic periods" (producer, member of a cocoa-coffee cooperative).

Thanks to producer awarness, the harm caused by climatic events can be limited. Heavy rain or drought can affect harvests, for instance, by reducing the quantities produced. Such a reduction in the quantities produced causes a decline in the sales revenues and margins of producers, leading to a long-term risk to employment: "We raise awareness of the consequences to workers by explaining that they could lose their jobs" (assistant to the head of the Fairtrade premium management committee, banana plantation).

The actions of producer organizations extend beyond Fairtrade-certified producers to raise awareness among producers in non-certified or conventional plantations which are located near certified agricultural parcels: "Raising the awareness of neighbors involves explaining what pesticides can cause. There is a three-month training period with our producers and neighboring producers" (assistant in charge of relationships between producers and customers, honey producer organization); "Measures are taken for producers that have Fairtrade products alongside organic ones; there are buffer zones there" (head of development and supply, coffee-cocoa producer and export organization).

Furthermore, meetings are organized at the national level; these make it possible to identify the needs of producers committed to certification and to discuss how to implement suitable actions. This occurred in October 2018 during the first National

Fair Trade Days, which were held in Abidjan (Le Point 2018). It was pointed out that the loss of forests had led to climate change, with a shorter, irregular rainy season and a longer dry season. In addition, this meeting emphasized the importance of moving towards a trade that tackles child labor and ensures children are sent to school (Le Point 2018), supporting small producers in a Fairtrade production process, paying them better and helping them to sell their raw materials in line with fairtrade principles (Devy Malu-Malu 2017).

11.3.2.4. *Measures to develop responsible consumption*

Beyond support actions centered on production, the efforts of producer organizations are focused on promoting the local consumption of fairtrade products. The local market is, in fact, still relatively undeveloped. According to Franck Koman, coordinator of the Ivorian Fair Trade Network, “half of fairtrade agricultural production is not sold due to a lack of strong local demand or suitable logistics, such as the poor state of roads”. Moreover, for Joël Bagbila, African head of Fair for Life, a certification body, “local consumption is almost zero”. He mentions that “people are not really concerned with quality so much as price”. For this professional, “we should first target local niche markets before attempting to reach the general public, as there are consumers who are willing to pay and who want the quality but do not know where to find it. In Burkina Faso, organic mangos, for example, are in very high demand, but no one knows where to buy them”.

According to Franck Koman, there is an effort to be made in Côte d’Ivoire in terms of communication to raise the awareness of local traders. One of the actions carried out in this regard is the organization of the first National Fair Trade Days, intended to inform farmers, consumers and the authorities of the benefits of fairtrade. The authorities should get further involved in the promotion of fairtrade (Le Point 2018). In this vein, the president of the *Artisans du Monde* association in Nice, Bruno Georges, in an interview given in November 2019 (text Box 11.1), highlighted the importance of the intervention of national and international bodies for guiding consumers towards fairtrade products.

Created in 1974, *Artisans du Monde* is a non-profit, activist supply network that advocates for an engaged approach to fairtrade. The network includes 130 local associations with 110 related stores through France and 5,000 volunteers. In 2017, the turnover of the active members of *Artisans du Monde* was 6.65 million euros (Artisans du Monde 2020).

Is there one label that particularly stands out among the various fairtrade labels?

In supermarkets, it’s the Max Havelaar label that stands out. It’s a simple certification that’s easy to understand for non-specialists. For us (*Artisans du Monde*), it’s

WFTO that stands out. Then, SPP (*Symbole de petits producteurs*) is developing more and more.

To determine the quality and requirements of each label, you have to look at the studies that have been conducted, such as the "*Guide des labels*" (2015), which was updated in 2018.

Do you think the Fairtrade label improves the living conditions of producers in the south?

It's much better! Obviously, there's always fraud; it's the same problem as in the organic field organics. At *Artisans du Monde*, we have progress criteria to improve conditions further; criteria such as the payment of a fair price or respect for human rights at work can be adjusted. It is therefore progress criteria that can be adapted to the economic, social and cultural situation of a given country that fairtrade organizations need.

Are there actions carried out by supporters of the Fairtrade label to reduce the harmful impacts of global warming on the countries where there are Fairtrade-certified producer organizations?

Yes, for all certifications, by promoting, for example, agroecology in coffee, cocoa, etc. I was able to verify this when I visited Bolivia and met producers; awareness-raising and training is being carried out in relation to these practices.

Does the Fairtrade label respond to consumer expectations in your opinion?

Yes; when we purchase chocolate that isn't labeled, we are ignoring everything that has gone into it. Yes, it responds to the demands of some consumers who want to consume better rather than more, and the environmental and social impacts are highlighted in our stores to increase awareness and turn this into an act of responsible consumption.

How can consumer concerns about the label be addressed?

There is always a part of the population that cannot buy these products, that is counting every last euro. People are starting to become more aware of local production. Fairtrade, for some, means products from the south that are therefore not environmentally friendly and do not have a direct impact on our population. We must change this perception. Transportation is only a part of the environmental footprint and any deterioration of working conditions on the other side of the world affects our workers too. People should understand and be convinced that their actions have an impact not only on producers, but also on the environment. Also, broader support at the governmental and European levels is needed.

Box 11.1. *Interview with Bruno Georges, President of the Artisans du Monde association in Nice*

11.4. Conclusion

The adoption of the Fairtrade label is synonymous with the establishment of sustainable production and the development of responsible consumption, despite the potential difficulties on the ground, such as climate issues, land management, the uniting of producers and the mistrust of consumers.

Our empirical study on Côte d'Ivoire enabled us to identify the support actions implemented by the branch of Fairtrade International in order to help producer organizations. These actions had three aspects: a social aspect (assistance provided to producers and their families, in particular); an economic aspect (with a guaranteed minimum price, reduced certification costs); and an environmental aspect (through training on improved use of resources and soil protection). All of these actions are intended to strengthen the application of fairtrade standards in relation to sustainable production.

Other actions are carried out by national bodies, such as meetings on fairtrade, not only to enhance sustainable production, but also to encourage responsible and local consumption. This can be facilitated by increasing the accessibility of fairtrade products for populations in the South, adapting prices to take local purchasing power into account, disseminating information in stores, and providing a guarantee that the environmental, social and the intrinsic quality of the products is greater.

Furthermore, professionals should help to alleviate doubts around compliance with standards and the effectiveness of the fairtrade system. Communication with consumers can thus be carried out in order to reassure them. This involves providing Ivorian consumers with information on the producer and product sold. This information must be mentioned on the product's packaging and delivered during tastings organized in supermarkets, or even via TV advertising. Interviews with committed producers should also be disseminated during national meetings or through televised cultural events.

Moreover, it is possible to integrate social and environmental objectives into marketing programs of fairtrade products. Another solution is to strengthen the guarantees and readability of fairtrade labels by establishing strict controls and by using a clear logo (GILCE 2020). Lastly, state support, at both the national and international levels, could increase the awareness of citizens.

11.5. References

Balineau, G. and Dufeu, I. (2012). Le système Fairtrade : une garantie pour les consommateurs ? *Mondes en développement*, 4(160), 11–26.

Castaldo, S., Perrini, F., Misani, N., Tencati, A. (2009). The missing link between corporate social responsibility and consumer trust: The case of fairtrade products. *Journal of Business Ethics*, 84(1), 1–15.

Clairouin, O., Foucart, S., Boyer, M., Bellanger, E. (2014). Comprendre le réchauffement climatique en 4 minutes [Online]. Available at: https://www.lemonde.fr/planete/video/2014/09/23/comprendre-le-rechauffement-climatique-en-4-minutes_4492721_3244.html [Accessed between September and November 2019].

Coulibaly-Ballet, M. (2019). Engagement éthique et certification sur le marché du commerce équitable : le cas de la certification Fairtrade en Côte d'Ivoire. *Revue éthique publique*, Special edition, 21(1).

De Pelsmacker, P. and Janssens, W. (2007). A model for fairtrade buying behaviour: The role of perceived quantity and quality of information and of product-specific attitudes. *Journal of Business Ethics*, 75(4), 361–380.

Dekhili, S., Coulibaly, M., Dufeu, I. (2017). Les consommateurs perçoivent-ils le prix des produits du commerce équitable comme étant juste ? *Décisions Marketing*, 85, 83–100.

Devy Malu-Malu, M. (2017). Afrique : cacao équitable cherche consommateurs désespérément [Online]. Available at: https://www.lepoint.fr/economie/afrique-consommation-cacao-equitable-cherche-consommateurs-desesperement-02-11-2017-2169399_28.php [Accessed 19 November 2019].

Diaz-Pedregal, V. (2007). *Le commerce équitable dans la France contemporaine : idéologies et pratiques*. L'Harmattan, Paris.

Dubreuil, E. (2020). Le commerce équitable, une autre économie pour demain. *20ème édition de la Quinzaine du commerce équitable*, 20 May.

Fairtrade (2020). Fairtrade Together: Covid-19 Briefing [Online]. Available at: https://www.fairtrade.net/news/fairtrade-together-covid19-briefing-on-27-april-2020 [Accessed 4 May 2020].

FLOCERT (2019). Votre route vers la certification Fairtrade [Online]. Available at: https://www.flocert.net/fr/solutions/ressources-fairtrade/ [Accessed between September and November 2019].

Guide International des Labels du Commerce Équitable "GILCE" (2020). © Commerce Équitable France [Online]. Available at: https://www.commercequitable.org.

Le Point (2018). Côte d'Ivoire : cet immense potentiel de commerce équitable si mal exploité [Online]. Available at: https://www.lepoint.fr/economie/cote-d-ivoire-cet-immense-potentiel-de-commerce-equitable-si-mal-exploite-27-10-2018-2266501_28.php [Accessed 19 November 2019].

Louppe, A. (2006). Contribution du marketing au développement durable. *Revue française du marketing*, 208, 7–31.

Makita, R. (2016). A role of Fair Trade Certification for environmental sustainability. *Journal of Agricultural Environmental Ethics*, 29(2), 185–201.

Max Havelaar (2019). Les standards Fairtrade [Online]. Available at: https://www.maxhavelaar.ch/fr/was-ist-fairtrade/fairtrade-standards.html [Accessed in September and November 2019].

Max Havelaar France (2020). Des évaluations pour mesurer et améliorer l'impact du commerce équitable [Online]. Available at: https://maxhavelaarfrance.org/le-commerce-equitable/impacts-et-changements [Accessed 23 May 2020].

Michrafy, M. (2013). Étude des effets des labels du commerce équitable sur le consentement à payer des consommateurs : application au cas du chocolat. PhD thesis, University of Montesquieu-Bordeaux IV, Pessac.

Radio Télévision belge francophone (2019). "Pas de droits, pas de chocolat !" : Oxfam veut un commerce du cacao plus équitable et écologique que le chocolonialisme [Online]. Available at: https://www.rtbf.be/info/societe/detail_pas-de-droits-pas-de-chocolat-oxfam-veut-un-commerce-du-cacao-plus-equitable-et-ecologique-que-le-chocolonialisme-actuel?id=10365619 [Accessed 15 November 2019].

Ruf, F., Leitz, E.U., Gboko, K.C., Carimentrand, A. (2019). Des certifications inutiles ? Les relations asymétriques entre coopératives, labels et cacaocultures en Côte d'Ivoire. *Revue internationale des études du développement*, 4(240), 31–61.

Sirdey, N. (2014). La durabilité sociale dans les standards de commerce équitable : une analyse de trois cas. *Journal des étudiants en développement international*, 3(1), 31–44.

Vimard, P. and Fassassi, R. (2011). Démographie et développement en Afrique : éléments rétrospectifs et prospectifs. *Cahiers québécois de démographie*, 40(2), 331–364.

12

Mobile Apps and Environmentally Friendly Consumption: Typology, Mechanisms and Limitations

Adeline OCHS[1] **and Julien SCHMITT**[2]

[1] *Audencia Business School, Nantes, France*

[2] *ESCP Business School, Paris, France*

12.1. Introduction

Numerous consumer studies, in France and internationally, show an increasing awareness of social and environmental challenges. Even more people are displaying an increasing willingness to adopt consumption behavior that preserves the natural environment (Obsoco 2019). Environmentally friendly consumption refers to "the use of goods and services that respond to basic needs and bring a better quality of life, while minimizing the use of natural resources, so as to not jeopardize the needs of future generations"[1]. Here, the notion of "use" is taken in its broad sense and includes not only the actual consumption of products and services, but also the stages that precede (e.g. researching information or purchasing decisions) and follow consumption (e.g. recycling).

Beyond these changes in environmental concerns, post-modern society is also characterized by the digital era that is transforming the relationships that consumers

1 Oslo Symposium on Sustainable Consumption (1994).

Marketing for Sustainable Development,
coordinated by Sihem DEKHILI. © ISTE Ltd 2021.

develop with consumption objects and social relationships. Consumers are now hyperconnected, especially through their cell phones, and continuously combine digital and physical channels to reach purchasing decisions. Mobile apps, which can act as "life coaches", differ from other digital devices because of their "mobiquity"[2], being both mobile and ubiquitous (i.e. omnipresent) (Badot and Moreno 2016; Badot *et al.* 2017). Being as close as possible to the situational context, mobile apps free individuals from spatiotemporal limitations. Their use is sometimes referred to as "ATAWADAC" (anytime, anywhere, any device, any content). This dimension opens up an enormous field of usage in consumers' lives. Specifically, numerous mobile apps are intended to help consumers to adopt more environmentally responsible consumption habits in their daily routines. Known as service-based or "servware", these mobile apps support consumers in carrying out one or several tasks without a commercial sales objective.

The aim of this chapter is to understand the role of service-based mobile apps in promoting environmentally friendly consumption. First, we provide an overview of the main mobile apps associated with environmentally friendly consumption, and we analyze their role within the consumer decision-making process (section 12.2). Then we describe the cognitive, social and emotional processes through which mobile apps influence consumption behavior (section 12.3). Finally, section 12.4 explores the challenges that emerge from these new apps and consumption modes that must be faced by the different actors in responsible consumption, namely brands, legislators and mobile apps themselves.

12.2. A typology of environmentally friendly mobile apps

12.2.1. *Environmentally friendly consumption and mobile apps*

There are numerous forms of environmentally friendly consumption. Using existing research (Daniel and Sirieix 2012), we identify and categorize different environmentally friendly consumption practices. Table 12.1 provides an overview of these pratices.

Mobile apps support a wide range of environmentally friendly consumption practices. They can be devoted to helping to reduce consumption amounts (e.g. 90 jours, WAG, Cleanfox, ECA), recognizing sustainable brands and companies (e.g. BuyOrNot, Good On You), choosing environmentally friendly products (e.g. Yuka, QuelCosmetic, Ocean's Zero), selecting distribution chains (e.g. Etiquettable, La Ruche qui dit Oui), or adopting optimal recycling methods (e.g. Geev, Too Good to Go, Frigo Magic).

2 Offering consumers the possibility to order at any time, irrespective of where they are (Badot *et al.* 2021).

Fields	Environmentally friendly consumption practices
Amount of consumption	Deconsumption: reducing quantity Need relevance: is the product really necessary?
Brand/company	Choose environmentally commited brands/companies Environmental commitment can take different forms: animal welfare, protection of animal species, waste reduction, etc. Indicated on labels or in the company's mission statement
Products/services	Choose environmentally friendly products Labeled Local Seasonal Low energy consumption Low pollution Decommissioned With little or no packaging
Distribution	Buy locally and/or without intermediaries Favor local and independent retailers
Mode of consumption	Adopt daily environmentally friendly practices in the private sphere Own-made production Opt for public transport or soft mobility Reduce consumption of energy, water, etc.
Mode of recycling	Encourage recycling Sort waste Give away or exchange products Re-market products

Table 12.1. *Types of environmentally friendly consumption practices*

To encourage these different practices, environmentally friendly mobile apps use different types of activity (Brauer *et al.* 2016):

1) Inform: provide relevant, geolocalized, personalized and adapted information to allow consumers to quickly adjust their behavior to different situations.

2) Educate: engage consumers in a learning process by explaining the reasons behind specific environmental choices.

3) Play: develop a playful dimension through games, quizzes and competitions.

4) Collaborate: encourage consumers to share their ideas and best practices with the community. This can refer to the mobile app's community (e.g. Etiquettable, see Box 12.1) or a community on another social network.

Figure 12.1. *Use of mobile apps in consumption (source: @MariannePetit)*

A collaborative mobile app

Etiquettable aims to allow people to consume differently by providing relevant information about fruits and vegetables, endangered/protected species of fish, zero waste tips, recipe ideas, lists of sales outlets, and commited restaurants. The information provided by the app comes from different scientific sources (Environment and Energy Management Agency, National Agency for Food, Environmental and Occupational Health Safety), non-governmental organizations (Ethic Ocean) and other contributors (chefs, nutritionists, etc.).

Each user is invited to participate on the app by giving suggestions for meals, places and ways to fulfil specific needs. The collaborative dimension is essential as it enables the app to adapt to user needs. It also promotes the involvement of users, either as co-creators of content or as mere readers of other members' comments.

Provide valuable information during the pre-purchase stage

Today, consumers face significant difficulty in finding correct information, as the messages they receive are often vague and are sometimes subject to controversy. With nearly 200,000 users, we have 300 daily hits, with an average connection duration of 2.5 minutes. The app steps in before the purchasing decision, by helping users in their search for information at all stages of the decision-making process: the preparation for shopping, the choice of stores and restaurants, the cooking or simply in reflection upon one's consumption behavior.

In the future: entertainment

In the future, we would like to develop more entertaining challenges. The app could assess the impact of potential changes in consumers' behavior, in order to increase their motivation for change. We would also like to encourage the development of the eco-score.

Box 12.1. *Interview with Marianne Petit and Shafik Asal, Founders of Etiquettable*

12.2.2. *The different stages of the purchase decision-making process of environmentally friendly products*

Stages of the decision-making process	Description of the stage	Application within environmentally friendly consumption
1) Need recognition	Tension between an actual state and an ideal state	Environmental utopia Leaning towards a more environmentally friendly ideal
2) Search for information	Aggregation of memory data in one's environment	Critical phase in behavioral change Issues of information overload and misinformation
3) Evaluation of alternatives	Selection of product sets	The ecological criterion should belong to the most influential criteria Issue of apps pre-prioritizing criteria
4) Purchase decision	Selection of the product to be bought, as well as its payment, delivery and/or transportation	Delivery and/or transport: a key issue in environmentally friendly behaviors
5) Post-purchase	Product usage, recycling, product's evaluation and recommendation	Use: increase the product's lifespan Recycling: donation, exchange, resale, waste reduction, etc. Evaluation: recommendations between consumers

Table 12.2. *Stages of the decision-making process and implications for environmentally friendly consumption*

The consumer decision-making process is often broken down into five stages, from need recognition to the post-purchase stage (see Table 12.2). In this section, we are going to analyze environmentally friendly consumption through the prism of this model and clarify the role of mobile apps at each of these stages.

The *need recognition* stage is defined as a state of tension resulting from a discrepancy between an actual state and an ideal state. The consumer then looks for potential ways to relieve this tension. Need recognition can result from a societal evolution: this is the case with the climate emergency, which has provoked a strong collective ecological awareness. A tension arises among consumers when they are confronted with the gap between their current (not necessarily environmentally friendly) behavior and their rising ecological concerns. For example, consumers may be willing to limit the amount of waste they generate but also fail in transforming this willingness into actual behavioral changes. The need recognized here is the will to find a way to adopt more environmentally friendly practices. Downloading environmentally friendly mobile apps is the first step toward the fulfilment of this need, as consumers acquire the tools that will allow them to change their behavior.

Once a need has been recognized, consumers engage in a search for information. Their aim is to get sufficient knowledge to make a decision that they consider appropriate. For everyday purchases, the search for information is restricted to the consumer's own memory. For more involved purchases, however, as is generally the case with environmentally friendly consumption, consumers tend to look for information in their external environment, by discussing matters with their peers (family, friends, or other consumers) and consulting different physical media (e.g. packaging, advertising) or online sources (e.g., social networks, websites, or blogs). Indeed, most consumers have little knowledge about environmental issues. In their quest for information, consumers quickly become drawn into an ocean of environmental claims conveyed by brands ("information overload"), especially as they often question the credibility of these arguments ("misinformation"). Is the argument put forward on the packaging really an environmental one? Is it greenwashing? Numerous service-based mobile apps are focused on this key stage of sharing and conveying appropriate information. They aim to provide objective and simplified information at the very moment when the decision is made, in order to counter the effects of information overload and misinformation (see Box 12.1).

The *evaluation of alternatives* by the consumer consists of selecting a set of products that would be suitable for purchase ("the consideration set"), according to some personally relevant criteria. These criteria depend on the costs and benefits associated with the product, such as comfort, price or esthetic. The environmentally friendly criterion belongs to the evaluation criteria. The adoption of environmentally friendly consumption behavior presupposes that this criterion significantly influences the product choice. "Scanning" apps (such as Yuka, Quel Cosmetic, BuyOrNot) assist consumers in this stage by scoring and ranking the different alternatives according to predefined criteria. This facilitates consumer decisions. Mobile apps provide simple assessments of products, broken down into three or four categories of criteria, based on the analysis of a large amount of data.

The *purchase decision* includes not only the final choice of the product bought, but also its payment and choice of its delivery and/or transportation mode. This is important as the different modes of delivery (at home or in store) differ in terms of environmental impact. Local purchases can have an environmental advantage if the means of transport are adapted and the logistics are optimized. If consumer trips to different stores are multiplied when buying local products, or if producers travel longer distances with lower quantities, the environmental gain of local options will be less. *La Ruche qui dit Oui* app, which promotes local producers, plays a crucial role during this stage, even though its offer remains limited because of the low number of available options. More broadly, apps such as *90 jours* or *Ocean's Zero* organize environmental challenges and thereby increase consumer awareness during the purchasing decision stage. The *90 jours* app, for instance, offers the "I joined the CSA (Community Supported Agriculture)" challenge, which promotes local producers.

The *post-purchase* stage includes three dimensions that have important implications in terms of environmentally friendly consumption. The first dimension is the *use of the product*: the way consumers use a product (e.g. how they wash it, whether they repair it) has an impact on its longevity. The Energy Consumption Analyzer app falls into this category by analyzing home energy consumption in order to reduce it. The second dimension is the *end of life of the product*, which is directly related to waste reduction (Guichard 2019). Rather than discarding the product in an ordinary trashcan, consumers must wonder whether it can be donated, recycled, exchanged, resold or reused in another way. The Geev app encourages products' second lives thanks to a system of donations between consumers. The WAG app offers a series of challenges to lean towards a zero-waste approach. The Frigo Magic app is focused on home cooking and offers recipes that use leftovers, thereby avoiding waste. The final dimension involves the *evaluation of products* (comments, opinions, notes). In a collaborative approach using virtual links, consumers communicate with other consumers to advocate more environmentally friendly practices or denounce environmentally harmful products. The degree of trust placed in the information and opinions of other consumers is greater than when the information originates from a private company.

Table 12.3 provides an overview of the role of mobile apps at the different stages of purchasing decisions.

Field (see Table 12.1)	MOBILE APPS			STAGES OF PURCHASING DECISIONS				
	App name	Number of downloads (Google Play)	Description	1) Need	2) Search for information	3) Evaluation of alternatives	4) Purchasing decision	5) Post-purchase (use, recycling, evaluation)
Level of consumption	Energy Consumption Analyzer	Between 100,000 and 500,000	Analysis of home energy consumption					*Use* Inform
Level of consumption	Cleanfox	Between 500,000 and 1 million	Unsubscribes and deletes emails and newsletters					*Recycling* Inform
Distribution	La Ruche qui dit Oui	Between 50,000 and 100,000	Application enabling purchases to made directly from producers				Inform	
Mode of consumption	Yuka	> 10 million	Scanner analyzing the nutritional composition of food products			Inform		
Mode of consumption	Good on You	Between 100,000 and 500,000	Scans fashion brands according to 3 criteria: environmental, animal and human			Inform		

Fields (see Table 12.1)	MOBILE APPS			STAGES OF PURCHASING DECISIONS				
	App name	Number of downloads (Google Play)	Description	1) Need	2) Search for information	3) Evaluation of alternatives	4) Purchasing decision	5) Post-purchase (use, recycling, evaluation)
Mode of consumption (continued)	Quelcosmetic & Inci Beauty	Between 500,000 and 1 million	Scans beauty products to analyze their composition			Inform		
	90 jours	Between 500,000 and 1 million	Environmental challenges in line with user profiles	Play		Play	Play	*Recycling/ Use* Collaborate
	WAG	Between 100,000 and 500,000	Challenges/coaching to reduce the user's carbon footprint		Educate/ Inform	Educate Collaborate	Educate Collaborate	*Recycling* Educate/ Collaborate
	ODD en action	Between 50,000 and 100,000	Presents 17 sustainable development goals, ways of achieving them and proposed actions		Inform			*Evaluate* Collaborate

	MOBILE APPS			STAGES OF PURCHASING DECISIONS				
Fields (see Table 12.1)	App name	Number of downloads (Google Play)	Description	1) Need	2) Search for information	3) Evaluation of alternatives	4) Purchasing decision	5) Post-purchase (use, recycling, evaluation)
Mode of consumption (suite)	Defendeur	Between 50,000 and 100,000	Games to learn about environmental protection (waste sorting, etc.)		Play			
Brands/companies Mode of consumption	BuyOrNot	Between 100,000 and 500,000	Scans products to determine their ethical impact and their nutritional composition		Inform	Inform		
Product/service Distribution Mode of consumption	Etiquettable	Between 100,000 and 500,000	Collaborative application providing recipes, ingredients, restaurants, sustainable producers and tips for sustainable consumption		Inform Educate			*Recycling* Inform

Fields (see Table 12.1)	MOBILE APPS			STAGES OF PURCHASING DECISIONS				
	App name	Number of downloads (Google Play)	Description	1) Need	2) Search for information	3) Evaluation of alternatives	4) Purchasing decision	5) Post-purchase (use, recycling, evaluation)
Mode of recycling	Too good to go	> 10 millions	List of local stores selling the day's unsold produce		Inform			
	Frigo Magic	Between 500,000 and 1 million	Ideas for simple and cheap recipes using leftovers		Inform			
	Geev	> 1 million	Platform for donations between private users (objects and food)					*Recycling* Collaborate/ Inform
	Guide du Tri	Between 100,000 and 500,000	Scan of products to inform users how to sort waste					*Recycling* Inform
	Ocean's Zero	Between 50,000 and 100,000	25 challenges to reduce waste and protect oceans	Play and educate	Play and educate	Play and educate	Play and educate	

Table 12.3. *Environmentally friendly mobile apps and purchasing decisions*

12.3. The influence of mobile apps on behavior

While consumers claim to be ever more aware of social and environmental issues, these environmentally friendly values are only reflected to a limited extent in their purchasing behaviors. This gap between values and behaviors has been widely documented in research. A consequence of this phenomenon, often referred to as the 'green gap' or 'action-value gap' (Auger and Devinney 2007), is that it is more efficient to directly influence behaviors than to try to influence values to promote environmentally responsible consumption. In this section, we show how mobile apps contribute to reducing the green gap by influencing purchasing decisions in three ways, cognitively, socially and emotionally (see Figure 12.2).

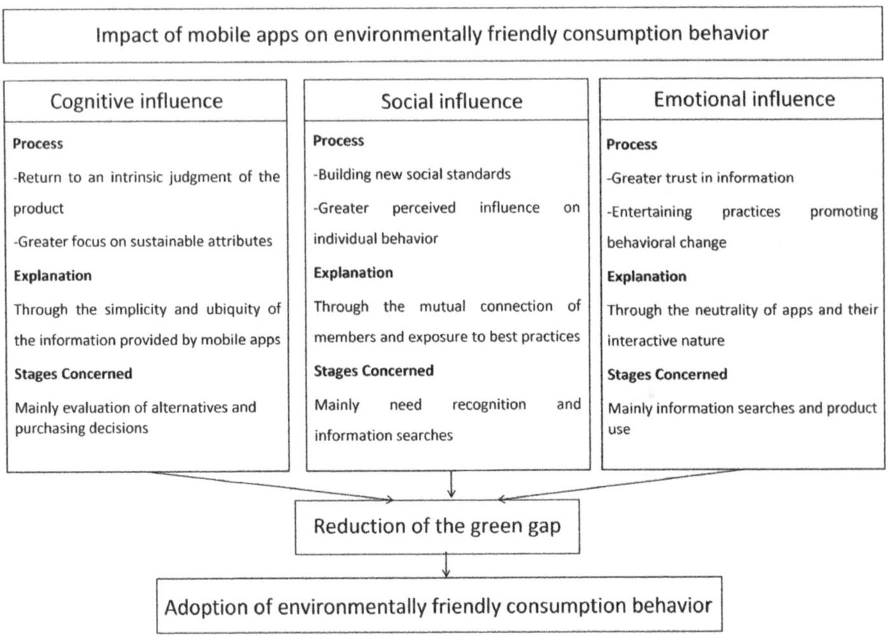

Figure 12.2. *The influence of mobile apps on environmentally friendly consumption*

12.3.1. *The cognitive influence of mobile apps*

Mobile apps contribute to reducing the green gap thanks to their capacity to reorient consumer decisions towards environmental attributes at the time of purchase. Consumers evaluate products according to different types of criteria (Langan *et al.* 2017). Some criteria are related to the product's intrinsic attributes, i.e. how it was made, its composition or its weight. Consumers rarely have sufficient expertise to form an accurate evaluation of products on the basis of these intrinsic attributes. Instead, they rely on extrinsic criteria unrelated to the product itself. Consumers thus assess the product indirectly, using cues such as brand name, brand positioning, perceived price or merchandising intended to draw attention to the product. These marketing tools have acquired such an influence on consumer decisions that they have sometimes come to replace intrinsic judgments completely. This phenomenon contributes to explaining why consumers do not choose products in accordance with their environmental characteristics, even if they consider such characteristics important.

Ecolabels have long been a visual way of providing information about products' intrinsic attributes: how they are made (i.e. organic agriculture) or what their ingredients are (i.e. palm oil-free). Mobile apps go even further. Thanks to their ability to present information with great simplicity, using language and cues that are immediately understood by consumers, mobile apps give consumers simple and immediate access to intrinsic attributes, even to consumers with a low level of expertise. For instance, the app BuyOrNot provides access to a highly simplified visual description of products' compositions, and their green/yellow/red color-coded presentation is immediately and easily interpretable for consumers.

The ubiquitous nature of mobile apps also represents a key element. Indeed, not only are intrinsic criteria made more comprehensible to consumers, but, in addition, mobile apps allow consumers to focus on these intrinsic criteria at any time, including at the moment of purchase. Generally, at the moment of purchase, consumers are more focused on more visible criteria, such as the price, which are generally closer to their personal interests (e.g. financial interests) and therefore have a greater influence on the decision (Olson 2015). By focusing the decision on environmental, social or nutritional criteria, mobile apps allow intrinsic and environmentally friendly criteria to play a more important role in the decision-making process.

12.3.2. *The social influence of mobile apps*

The positive influence of mobile apps on environmentally friendly consumption also involves a mechanism of social influence. Indeed, the gap between pro-environmental attitudes and consumption behaviors also originates from social pressure: consumers are influenced in their decisions by their perceptions of what others think or do in a particular situation. Social standards sometimes influence purchasing behavior to a greater extent than individual attitudes. Several studies show that consumers are more inclined to adopt environmentally friendly behaviors (e.g. recycling waste or minimizing energy use) when they are made aware that this behavior has been adopted by many other people, rather than when the environmental benefit of this behavior is explained to them (De Jong *et al.* 2017).

Mobile apps are interesting in that regard: they situate the act of consumption in a social context by showing users the behaviors that have been adopted by the other members of the community. For example, the Too Good to Go app allows different actors to connect with each other to optimize waste management, and therefore creates new social standards for app users.

The use of social influence also enables another inhibitor of environmentally friendly consumption, namely the "perceived behavioral impact" (Leary *et al.* 2014). Consumers tend to believe that they have a low behavioral impact, meaning that their individual actions taken in isolation only have a limited impact on the world: "What does it change if I have a bath rather than a shower, compared to all the water used in big companies' factories?". Mobile apps allow for a better awareness of one's own behavioral impact. Apps such as Geev allow consumers to easily visualize collective performance, such as the waste avoided thanks to the behavior of all of the users of the app combined, or the amount of products that will not need to be produced thanks to this gain (see Box 12.2). The WAG app offers environmental challenges to allow users to tackle environmental issues at their own level. This helps consumers to incoporate environmentally friendly behaviors into their daily routines. For each of these challenges, WAG provides information about the participation of other users, thereby creating a social norm effect.

12.3.3. *The emotional influence of mobile apps*

Mobile apps also influence consumers in an emotional way. Firstly, consumers tend to consider information from mobile apps more positively than information conveyed by commercial brands. Indeed, they lack trust in the social or environmental claims they see in advertisements. These messages are seen as having low levels of credibility and neutrality (Karniouchina *et al.* 2009)[3]. Conversely, mobile apps are perceived to be a new and relatively neutral and objective actor. Consumers therefore consider these apps to be trustworthy sources of information, even though ultimately, they know little about the actors operating these apps.

The emotional power of mobile apps also lies in their ability to create enjoyment and trigger an entertaining aspect of environmentally friendly consumption. Some apps, such as Too Good To Go, organize fun competitions to increase their members' motivation to adopt sustainable behaviors. Numerous Yuka users explain that scanning products also

3 Credibility: indicates the level of trust that consumers have in a specific source of information, depending on whether this source is considered honest or sufficiently authoritative. Perceived neutrality: indicates the extent to which consumers perceive the source of a message as not having an interest in the consumer's reaction to that message (for example, a financial interest).

has a fun side, which is part of the experience and creates the desire to use the app[4]. The apps 90 jours and Ocean's Zero also play with these entertaining aspects by organizing challenges and competitions, and providing users with emotional rewards.

What is the aim of Geev?

Geev allows users to connect with each other and to give and collect food items in their area. The app offers a real alternative to selling products, unnecessary storage and waste in all its forms.

How do people use the app?

At the end of January 2020, Geev had almost 2 million users. Over 270,000 donation offers are posted every month, and Geev has already enabled more than 3 million objects and food products to have a second life.

Users have many different motivations. They may give away items out of generosity or for their own personal interest by freeing up space in their homes. It is also an act showing commitment to the fight against waste. Finally, it represents a quick and convenient alternative to selling one's products. Some life events are particularly suited to donations, such as moving house or having a child. It's a real barometer of life's milestones!

What is the role of Geev in environmentally friendly consumption?

What do we do with the objects that we no longer use or the food that we aren't consuming? Geev makes it possible to answer this question through donations between individuals. It thus plays an environmental role by extending the lifespan of objects and reducing waste. The human factor is also very important as service is a vector for social connection: it allows users to meet other people living in the same town or neighborhood, irrespective of their social background. The app is therefore fully committed to the logic of the circular economy.

And in the future?

We want to normalize the use of donations between individuals and establish it in consumers' daily lives. After France and Canada, we want to roll out our app throughout Europe to make donation easy, practical and fast.

Box 12.2. *Geev, the app popularizing donations among individuals. Created in 2017 by Hakim Baka and Florian Blanc*

4 Quotes from a qualitative study based on interviews and in-store observations of 20 users of the Yuka app between January and May 2018 with the aim of understanding how the app alters their decision-making process.

12.4. What are the implications for the different actors in environmentally friendly consumption?

	Brands	Regulators	Mobile apps
Risks and challenges	– Loss of influence of well-known brands to the benefit of third-party actors, such as mobile apps	– Little regulation of mobile apps' content – Bypassing evaluation criteria	– Environmental impact of digital technology – Harmful social impact of use of smartphones
Solutions	– Greater transparency – Clearer positioning and communication	– Clearer legislative framework for apps – More comprehensive evaluation standards	– Management of "digital sobriety" – Limitation of the risk of smartphone dependency

Table 12.4. *Main challenges for the different actors in environmentally friendly consumption*

Sections 12.2 and 12.3 showed how mobile apps influence consumer behavior. In this section, we explain the challenges posed by this phenomenon for the different actors in environmentally friendly consumption, namely brands, public authorities and mobile apps themselves. Table 12.4 summarizes a general overview of these implications.

12.4.1. *At the brand level: (re)learning how to communicate*

12.4.1.1. *The risks: a loss of influence of brands on consumer decisions*

One of the reasons behind the success of mobile apps lies in the loss of consumer trust in big brands, regarding issues linked to environmentally friendly consumption. This leads to the development of third-party actors that are perceived to be more honest by consumers. First, there are the very new brands, offering a 100% responsible range, which have been able to gain consumer trust. This is case for brands such as Lush or Veja. Second, mobile apps, perceived as non-commercial actors working in the interest of the consumer, are also slowly becoming reference points for consumer decisions.

Some consumers, whose use of the Yuka app has been observed,[5] even go so far as denigrating brands. It is not uncommon to hear consumer quotes, such as "I like scanning products to see the trash they try to sell us" or even "I don't even look at brands anymore, I just look at what Yuka says". Even if such consumers are still a minority today, such quotes show that a significant change is occurring and that brands must prepare to respond by leading the conversation on these issues. What is the point of investing in brand creation and management if this attribute is no longer taken into account in purchasing decisions?

12.4.1.2. *A new communication style: transparency*

A new communication style from brands is required to move toward greater transparency. Transparency is a significant expectation among consumers and represents one of the reasons for the success of mobile apps. In his article "The world behind the product", de Leeuw (2008) explains the principle of product life cycle analysis. It involves the assessment of the product's environmental impact throughout its entire material life, from the extraction of raw materials to its end of life. Consumers expect such information to be clearly communicated to them, in order to be able to compare brands on the basis of such criteria. Research shows that advertisements that provide claims related to the product life cycle in a simple way are able to improve attitudes towards the brand and towards the ad, as well as the perceived credibility of the message (Bascoul *et al.* 2014). Brands that pioneer the use of such environmentally friendly communication strategies will be able to create a significant competitive advantage. For example, the Loom clothing brand has adopted a "100% transparency" strategy through which it communicates its key strengths about its environmentally friendly production process, as well as the areas where it still needs to improve.

12.4.1.3. *A new communication style: adopting clear positioning*

Positioning consists of ensuring that the brand occupies a unique place in the minds of target consumers. Brands will have to determine how to integrate sustainable development into their positioning. Do they want to use sustainability as a key point of differentiation, or as a mere secondary characteristic? The answers to these questions will depend on the competitive context and the expectations of the target audience. In any case, brands can no longer ignore this issue and must take a very clear stance. The chosen positioning should be backed up by proof. Such proof could be tangible cues related to elements of the product's life cycle, or indirect cues such as testimonials from third parties, including labels, NGOs or – why not – mobile apps. The Malongo coffee brand has opted for brand positioning that confers

5 Same study cited previously.

equal importance on two main elements: the excellence of the coffee and respect for small producers. Only the second element is related to environmental concerns; this is illustrated by visual "proof", such as testimonials from producers, environmentally friendly labels or even pictures of plantations, on an equal basis with symbols communicating the quality of the coffee.

12.4.2. *Much needed regulation*

12.4.2.1. *The regulation of mobile apps*

A challenge for regulators consists of verifying who the actors behind mobile apps are and ensuring the accuracy of the information they provide to consumers. Indeed, these actors are becoming key influencers of consumer decisions: it is therefore necessary to protect consumers by ensuring that they are exposed to trustworthy information. For instance, some initial questions have arisen about the information provided by apps such as Yuka. Its information about products comes from the Open Food Facts database, which does not guarantee its accuracy. Major retailers are trying to develop competitive apps and provide product information and evaluations that differ from Yuka's: they are perceived as too indulgent towards brands. At present, information provided by apps tends to be provided by the brands themselves, which raises questions about the necessity (and the possibility) for apps to verify this information.

12.4.2.2. *Bypassing the product evaluation criteria*

Brands respond to the assessment of their products by some apps by changing the composition of their products. In this way, they hope to obtain good evaluation scores. The danger is that the product manufacturers are tempted to replace additives that score poorly or added sugar with other equally harmful ingredients that would not penalize the product scores provided by apps. Public authorities must take this into account by providing precise information on the types of indicators that apps should consider. This is also an issue that the apps themselves must take seriously; they could risk losing their credibility if consumers were to start noticing the possibility of bypassing their criteria.

12.4.3. *Taking into account the potential negative effects of mobile apps*

12.4.3.1. *The environmental impact of mobile apps*

The energy consumption of mobile apps is substantial. The Shift Project (2020) indicates that the impact of digital technology accounts for 3.7% of global greenhouse gas emissions. With the ever increasing growth of digital technology, this proportion could rise to 8.5% by 2025. Smartphones, including their manufacturing and use stages, account for 11% of digital consumption. This proportion is also likely to rapidly increase because of the gowing importance of phones as part of the total time spent on digital consumption.

"Digital sobriety" is an emerging movement that advocates the development of more restrained digital services; it calls for moderation of the daily use of digital technology (Bordage 2019). Mobile apps that are intended to promote environmentally friendly consumption must take this issue into consideration. To be consistent with their objectives, they should attempt to minimize the environmental impact of their usage. This involves measures such as raising awareness of the environmental impact of digital technology, minimizing data transmission during app usage, minimizing the time needed to obtain optimal use of the app and, in the longer term, changing the reference indicators. Some newly developing apps, such as Plana, aim to help users to optimize the energy consumption of their phones.

12.4.3.2. *The social impact of mobile apps*

Mobile apps also have a social impact. The phenomenon of smartphone dependency has been widely documented in academic research (Shekhawat and Batra 2017). It is defined as an excessive and difficult to control need to use one's smartphone, causing a negative impact on the other areas of an individual's personal, social and professional lives. Social apps are particularly likely to cause such an addiction (Kwon *et al.* 2016). Indeed users tend to develop a social addiction to other people's reactions on mobile apps. The mere sound of a smartphone notification indicating a message or a "like" could cause an addictive reaction and, consequently, a deterioration of the consumer's well being.

Mobile apps that promote environmentally friendly consumption must take this phenomenon into account. Is it not the role of mobile apps to educate consumers on the potential harmful impacts of digital technology? Adopting environmentally friendly behaviors also involves environmentally friendly practices in digital consumption that must be driven by apps.

12.5. Conclusion

Many mobile apps offer consumers daily support to acquire and reinforce environmentally friendly consumption behavior. In this chapter, we have shown that these apps intervene at different stages of the purchasing decision and that thay play various roles, from providing simple and objective information to transforming behaviors by using "challenges" to change consumption habits. We have also analyzed the psychological and social mechanisms that explain how mobile apps influence environmentally friendly consumption: cognitive influence (a return to an intrinsic judgment of the product and a greater focus on attributes related to sustainable development); social influence (the creation of new social standards and a greater perceived influence on individual behavior); and emotional influence (a new and trustworthy source of information and entertaining ways of promoting behavioral change). Lastly, we have described the challenges that these new mobile apps pose for the various actors in environmentally friendly consumption, namely brands, regulators and mobile apps themselves. For brands, the major challenge lies in the growing loss of their influence over consumer decisions to third-party actors, such as mobile apps. Regulators must face challenges linked to the lack of regulations imposed on mobileapps, and the possibilities that brands have to bypass the criteria used by apps to assess their products. Lastly, mobile apps' main challenge lies in the proper management of their environmental and social impact.

The emergence of mobile apps is transforming the way consumers make their decisions. Specifically, mobile apps make it possible to break established routines that have a strong effect of inertia. Consumers must then learn how to adopt new behavioral routines, in which ethical and responsible considerations play a bigger part. Today, purchases made with the information provided by mobile apps are still perceived as too long by consumers: looking for products, scanning barcodes and comparing results are actions that significantly increase the time spent on shopping. Consumers do not want to devote so much time to their daily purchasing decisions. However, this time will gradually be reduced after an initial learning stage: with their first uses of mobile apps, consumers will slowly discover new environmentally friendly brands that will gradually replace their previous habits. Consumers will develop new decision-making routines by shifting their considerations towards products that are better rated by apps. It will be interesting to study the risk faced by brands reluctant to move toward more sustainable practices, the risk of gradually becoming neglected by consumers and removed from the consideration set. This would represent a tremendous risk, as the marketing investment to get back into the consideration set would be huge, and actually much higher than the investment necessary to actually become environmentally friendly. Such a perspective shows the need for brands to quickly take a stance on the sustainability issue.

12.6. References

Auger, P. and Devinney, T.M. (2007). Do what consumers say matter? The misalignment of preferences with unconstrained ethical intentions. *Journal of Business Ethics*, 76(4), 361–383.

Badot, O. and Moreno, D. (2016). *Commerce et urbanisme commercial. Les grands enjeux de demain : regards croisés*. EMS, Cormelles-le-Royal.

Badot, O., Lemoine, J.F., Ochs, A. (2021). *Distribution 4.0*. Pearson, Paris.

Bascoul, G., Schmitt, J., Rasolofoarison, D. (2014). Consumers' biased perceptions of the environmental impact of products, and their influence on advertising efficiency. In *The Sustainable Global Marketplace*, M. Conway Dato-on (ed.). Springer, Berlin.

Daniel, M. and Sirieix, L. (2012). Les pratiques durables : une forme de résistance ordinaire ? *Décisions Marketing*, 11–24.

De Jong, A., Schmitt, J., Schmidt, K. (2019). Do as I do, not as I think: Disentangling the impact of different types of social norms on customer environmental consumption. *The 18th International Marketing Trends Conference 2019*. Venice.

De Leeuw, B. (2008). The world behind the product. *Journal of Industrial Ecology*, 9(1/2), 7–10.

Guichard, V. (2019). *Du gaspillage à la sobriété : avoir moins et vivre mieux ?* De Boeck, Paris.

Karniouchina, E.V., Moore, W.L., Cooney K.J. (2009). Impact of mad money stock recommendations: Merging financial and marketing perspectives. *Journal of Marketing*, 73(6), 244–266.

Kwon, H.E., Hyunji, S., Sang, P.H., Wonseok, O. (2016). Excessive dependence on mobile social apps: A rational addiction perspective. *Information Systems Research*, 27(4), 919–939.

Langan, R., Besharat, A., Varki, S. (2017). The effect of review valence and variance on product evaluations: An examination of intrinsic and extrinsic cues. *International Journal of Research in Marketing*, 34, 414–429.

Leary, R.B., Vann, R.J., Mittelstaedt, J.D., Murphy, P.E., Sherry, J.F. (2014). Changing the marketplace one behavior at a time: Perceived marketplace influence and sustainable consumption. *Journal of Business Research*, 67, 1953–1958.

Obsoco (2019). L'Observatoire des perspectives utopiques [Online]. Available at: http://lobsoco.com/lobservatoire-des-perspectives-utopiques [Accessed October 2019].

Olson, E. (2015). It's not easy being green: The effects of attribute tradeoffs on green product preference and choice. *Journal of the Academy of Marketing Science*, 41(2), 171–184.

Shekhawat, H. and Batra, D. (2017). Psychological characteristics and mobile phone addiction among adolescents. *International Journal of Management & IT*, 15(2), 76–82.

The Shift Project (2020). Déployer la sobriété numérique [Online]. Available at: https://theshiftproject.org/wp-content/uploads/2020/01/2020-01.pdf [Accessed January 2020].

13

Digitalization in the Service of Socially Responsible Consumption? Focus on Food Consumption[1]

Christine GONZALEZ[1], Béatrice SIADOU-MARTIN[2] and Jean-Marc FERRANDI[3]

[1] *Argumans, University of Le Mans, France*

[2] *MRM, University of Montpellier, France*

[3] *LEMNA, ONIRIS, Nantes University, France*

13.1. Introduction

"We have run out of excuses and we are running out of time. We have come here to let you know that change is coming, whether you like it or not."[2] These were the words of Greta Thunberg, the young Swedish environmentalist, at the *24th Global Climate Conference*. While the need to change behavior in production and consumption to respond to current environmental challenges is recognized, it remains the case that actual behavioral change has not yet been seen.

1 This chapter has been written as part of the IP-Ag project (Perceived Intrusiveness of Anti-Waste Apps), financed by the Pays-de-la-Loire region and ADEME (French Environment and Energy Management Agency).

2 https://www.lemonde.fr/climat/video/2018/12/15/nous-sommes-a-court-d-excuses-le-puissant-discours-d-une-jeune-ecologiste-suedoise-a-la-cop24_5398255_1652612.html [Accessed December 13, 2019].

Indeed, in many areas, such as sorting waste, consuming environmentally friendly products or limiting food waste, consumers are developing an awareness of and a positive attitude towards responsible consumption, even as the observation of practices highlights the weakness of actual change. The marketing literature has widely addressed this attitude–behavior gap, as well as the information processing strategies that the consumer tries very hard to develop to justify it (Rodhain 2013; Séré de Lanauze and Siadou-Martin 2014). Intuitively, it seems that difficulties in adoption can involve two areas: information and action. In other words, are consumers informed about which behaviors they should avoid and which they should adopt? Are they able to do this themselves or through the means offered to them? What inhibits or facilitates responsible behaviors?

For François-Lecompte and Valette-Florence (2006), socially responsible consumption translates into a set of practices structured around five dimensions: organization, which leads to promoting and/or avoiding some distribution channels or products; purchasing "sharing products", which redefines the sharing of the value created during trade; small businesses; the origin of the product; and the volume of the product, which indicates consumer needs. Buying in second-hand clothes stores or flea markets, buying organic or fairtrade products, and not buying products from polluting companies are among the examples of socially responsible consumption practices.

At the same time as there is this awareness of the negative impacts of our current consumption system on the environment, consumers are becoming increasingly connected. For example, smartphones, which are now our main means of connection, are genuine *memory tools* or *everyday personal assistants*. A study carried out by BigBen Connect in 2017 specified that *many things in our lives are on our phones and not elsewhere*, such as our alarm clock, photos, bank accounts, GPS[3] and favorite series. These technological tools thus offer original solutions and can seem to be valuable assistants in adopting socially responsible behaviors. In the field of home automation, some apps allow users to indicate their arrival and start heating their homes from a distance. In health, connected watches help to make information that is difficult for people to find more prominent, such as the effort made and calories burned. While all areas of consumption are affected, this contribution is focused on food consumption, a choice justified by, on the one hand, its importance in household budgets and in the economic system and, on the other hand, by the establishment of entrenched habits owing to culture and education.

This chapter is structured into two parts. Firstly, we consider the compatibility of digitalization and sustainable food. We will then address how digitalization also

3 Global Positioning System: satellite-based geolocation system.

paves the way for interesting areas of reflection or action for particular target audiences and in specific contexts.

13.2. The paradoxes of digitalization and sustainable food

Taking an interest in food consumption is inherently complex owing to its multifaceted nature. Sirieix (1999, p. 41) explained that "responding to a concern both fundamental and repetitive, the consumption of food products can be routine or celebratory, individual or collective, utilitarian or ostentatious, making the act of food consumption neither solely rational, nor banal: food is indeed not only a requirement for life, but also a tool in the service of health, pleasure and physical appearance." These different challenges are also included in the definition of sustainable food by the United Nations Food and Agriculture Organization. Sustainable food is "protective and respectful of biodiversity and ecosystems, culturally acceptable, accessible, economically fair and affordable; nutritionally adequate, safe and healthy; while optimizing natural and human resources". Responding to all of these challenges simultaneously leads to a questioning of practices relating to purchases, usage and consumption. In other words, buying, storing, preserving, preparing, cooking and valuing the leftovers are practices that merit the attention of researchers and practitioners. In this context, the issue arises of the compatibility of sustainable food with digitalization.

13.2.1. *What compatibility is there between digitalization and sustainable food?*

Nowadays many technical and technological devices are intended to facilitate the adoption of more responsible consumption. This is the case of anti-waste or nutrition-based mobile apps, interactive terminals in stores or connected packaging. Beyond this technical aspect, digitalization echoes a reconstruction of both the product range and distribution practices put in place by manufacturers or chains, and the purchasing and consumption practices adopted by consumers. In ordinary food shopping, "it is less a question for individuals of replacing their manner of shopping with another "more digitalized" manner, than of broadening their horizons, that is, having alternative means that allow them to adjust their actions according to their perception of the situation" (Comino 2017, p. 375). This reconstruction of practices can be seen at three different levels:

– digitalization enables a more efficient management of information in the consumer–company relationship in two ways. Companies can collect data on consumers and consumers can access relevant information when they make decisions, following the example of the Yuka app, which provides nutritional information during purchases;

– digitalization questions the spatiotemporal framework of trade. To the market-based serendipity that enables the discovery of something unexpected by chance or through insight, digital tools offer a virtual serendipity. In effect, at the touch of a button, the consumer has a significant choice of providers, distribution formats, products, and means of delivery and payment. However, this serendipity can conflict with modes of supply and the requirements of responsible food. For example, ordering directly from farmers or limiting food waste leads to meal planning and the consumption of leftovers from meals, and therefore to the consumer becoming more organized;

– moreover, the temporal dimension of consumption is transformed by the extension of the consumption experience and by interacting with other consumers, which takes the concrete form of posting messages or videos on social media. Through these different means, other consumers can become part of the private lives of the posters.

Questions around the frequency and nature of the intrusion into the private lives of consumers through notifications or data collection lead this debate towards ethical considerations: does the adoption of responsible behaviors justify all of the means, including the intrusive means?

The adoption of an anti-waste mobile app leads to the perceived "technical" characteristics of apps, such as ease, enjoyment, perceived interest, perceived informative nature, perceived savings and personalization of the app. These different aspects are shown in Table 13.1.

Furthermore, beyond these technical aspects, adoption also triggers a reaction to the behavior of the user of the mobile app: its intrusiveness, described in Table 13.2. More specifically, people express a feeling of intrusiveness: 1) aspirational intrusiveness; 2) invasive intrusiveness; 3) forced intrusiveness; 4) usurping intrusiveness; and 5) interfering intrusiveness (Gonzalez and Siadou-Martin 2019). A quantitative study[4] shows that the two main explanatory factors behind the intention to use or recommend a mobile app are linked to the perceived invasiveness and usurpation. Consumers would thus be particularly aware, on the one hand, of notifications and intrusion into their daily lives and, on the other hand, of the legitimacy of the company behind the app (Gonzalez *et al.* 2019).

4 Online study carried out in the fall of 2018 among a convenience sample of 145 French consumers. This questionnaire assessed the pre-use intrusiveness of a fictitious anti-waste mobile app.

Facilitators	Statement[5]
Perceived ease	Degree to which someone believes that the use of a particular system will entail no effort "A quick, easy to use app where you can easily see what you have in the fridge by first highlighting the products to use and also by offering recipes for the products you have in the fridge. There could maybe also be advice on how to arrange products in the fridge better to preserve them as well as possible, mainly for fruits and vegetables."
Perceived interest	Degree to which someone thinks that the use of this tool will improve their performance "I like app X as it can give us ideas for meals to avoid wasting what we have in the fridge. This can be a good thing for people who waste a lot, it could definitely help them to reduce their waste. The people likely to be interested in this are those who waste a lot, people who don't have a lot of money or big families because the apps allow them to see which stores have products that have reached their sell-by dates and that will therefore be cheaper."
Perceived enjoyment	Enjoyment that will be derived from use of a technology "It's fun and accessible to everyone but it can be annoying to always have to watch what you're doing, to always be looking at the app." "I'd like this to be fun and practical to use on a daily basis and to let me know when I have a product I need to eat soon."
Perceived informative nature	Information and advice that would be obtained on ways to avoid waste, as well as on environmental problems associated with waste "Er, my expectations of an app would be advice, for starters, on avoiding waste, info like "what is it that causes waste on a global scale", you know, er, poverty, people who don't have enough to eat, etc. Info like that and advice. And also recipes for food that's likely to be wasted. Yeah, actually, I'd say that'd be good if we know the consequences of the waste, because if we know them, it can push us to stop them."
Perceived savings	Possibility of making savings and obtaining reduced prices or good deals through the mobile app "What would the ideal anti-waste app be for you? Being informed of promotions relevant to my shopping list."
Personalization	Adaptation of the mobile app to behavioral, psychological and geographical elements relevant to the individual at a given time "A really useful app would be personalized by collecting loyalty points in a pot that can be reused in different stores with your smartphone."

Table 13.1. *Technical reasons for adopting an app*

5 These statements are from a qualitative study carried out in 2017 among a sample of 25 French consumers, composed of 8 men and 17 women. Its aim was to understand the reasons for adopting an anti-waste mobile app.

Dimensions	Definitions
Aspirational intrusiveness	Perception of intrusion into one's digital private life and of the collection of data on the consumer.
Invasive intrusiveness	Perception of intrusion into one's physical private life and of an interruption of the consumer's cognitive activities through notifications, for example, or through the invasion of the smartphone into the consumer's life.
Forced intrusiveness	Perception of intrusion into one's physical private life, people fear that their smartphones or their apps are becoming full of advertising, without their permission, or infected by malware that prevents them being easily used.
Usurping intrusiveness	Perception of intrusion into one's interactional private life, people relate the issue to the source, which may or may not be relevant.
Interfering intrusiveness	Perception of intrusion into one's psychological private life and that refers to "the perception by consumers of a control exercised by brands over their behaviors and their daily lives, and a restriction of their independence in decision-making and actions through the imposition of behaviors and beliefs, and ultimately an infantilization".

Table 13.2. *Dimensions of intrusiveness (source: (Gonzalez and Siadou-Martin 2019))*

13.2.2. *A critical look at consumer responsibilization*

The consumer is made aware of the challenges of sustainable food via the statements of various actors in the press and/or social networks. These actors can be companies (mobile apps focused on reducing food waste) or public actors (the "*Fabrique à menu*" ("make a menu") function offered by the site mangerbouger.fr as part of the French National Health and Nutrition Plan). They "train, guide and mold" the consumer (Cova and Cova 2009) by looking to modify "very intimate and identity-related practices, such as the preparation of meals or food shopping" (Gonzalez and Siadou-Martin 2019). In this regard, the question arises of the governmentality of the consumer (Stripple and Bulkeley 2017). Governmentality is based "on the concept that power has become decentralized and is exercised by multiple actors through various means; for example, private actors can be responsible through voluntary agreements, contracts, education, information campaigns and consultative practices" (Uggla and Soneryd 2016).

Governmentality is distinguished, on the one hand, by conceptualizations, discursive activities and knowledge, and, on the other hand, by "technologies", i.e. "mechanisms through which all kinds of authorities seek to construct, standardize and use the behavior, thoughts, decisions and aspirations of others to achieve objects they consider desirable" (Miller and Rose 2008). In other words, the common thread is to lead people to voluntarily establish the expected behaviors by shaping their behaviors and thoughts (Lovbrand and Stripple 2017). Giesler and Veresiu (2014), in particular, describe how institutions are making consumers responsible through four stages, as described in Table 13.3.

Stage	Definition of the stage
Personalization	This redefines the solution to a key social problem in terms of the development of a "morally enlightened" actor: responsible consumers differ from irresponsible consumers in their desires, aspirations and ability to make choices.
Authorization	This is based on economic data, psychological knowledge, etc. to justify the development and adoption of responsible behavior both legally and economically.
Empowerment	This develops a market (products and services) for ethical self-management.
Transformation	For consumers, this consists of adopting their new reformed "self-awareness".

Table 13.3. *The four stages of training a subject to be responsible (source: adapted from (Giesler and Veresiu 2014, p. 841))*

Digital tools contribute to an individualization of responsibility by giving rise to guilt in regard to social environmental issues and by proposing solutions to make consumption more virtuous/responsible (Fuentes and Sörum 2018). For example, anti-waste mobile apps allow consumers to be socially responsible, but in return guide and shape their behavior: in order to operate them, consumers must scan the barcodes of the products they buy. These apps can also ask consumers to share their ethical performances and to participate in data collection to provide the app with more information. Three areas arise from this that enable these tools to empower consumers: problematization, consumer agency and the configuration of ethical consumption. This is illustrated in Table 13.4.

Empowerment axis	Means of empowerment and example with mobile apps
Problematization	By highlighting the ethical dimension of consumption, mobile apps consider each act of consumption a dilemma or ethical problem.
	The platform "JeuxNeGaspillePas !", launched by the ANIA (French National Association of Food Industries) offers informative games and advice on "buying better", "storing better" and "consuming better".
Consumer agency	Mobile apps allow consumers to be actors in their ethical consumption by providing them with the information needed to make ethical decisions, making them aware of the complexity of consumer activities and, lastly, enabling them to position themselves as ethical consumers in relation to others.
	The mobile app Zérogachis offers consumers information on sales outlets where they can find products approaching their use-by date.
Configuration of ethical consumption	Mobile apps allow consumers to internalize ethical consumption by suggesting different actions they could take. To that end, they are based on a very large range of consumer figures beyond the ethical decision-maker, such as activists, for example.
	The mobile app FrigoMagic offers easy-to-cook recipes using products that will soon expire and are in the fridges of consumers.

Table 13.4. *Means of empowering actors (source: adapted from (Fuentes and Sörum 2018))*

13.2.3. *The environmental impact of digitalization*

There are serious potential environmental threats linked to new technologies throughout the three stages of their lifecycle:

– *the manufacturing of devices and their components*, which uses (sometimes rare) raw materials, as well as water, and leads to potentially polluting residue;

– *the use of the devices by individuals*, which translates into the electricity consumption of computers, as well as of data transfer or data storage systems, etc.;

– *the discarding of devices by consumers*, which can have harmful impacts on the environment (Murugesan 2008). In this regard, the International Energy Agency (IEA) refers to *energy waste* through the proliferation of connected objects that constantly consume energy[6]. The energy wasted was estimated at 80 billion dollars in 2013[7].

6 https://www.generation-nt.com/iea-rapport-appareil-electronique-internet-gaspillage-energie-actualite-1899762.html.

The use of technological devices to modify consumer behavior to be more responsible thus leads to significant tension between, on the one hand, "the negative environmental impact of producing, using and discarding the technological devices" and, on the other hand, the possibility of using them "to improve the environmental sustainability of companies and society" (Gholami *et al.* 2013).

This debate is also taking place among practitioners. Pierre Bonzom, CEO of ELA Innovation, confirmed in an article entitled "Connected objects: from an economic El Dorado to an environmental fiasco?" that "this approach will lead to greater production of electrical energy, needed to supply all of the cells and batteries used, as well an increase in the waves emitted, potentially even leading to a complete saturation". Going further, he specified:

> "The environmental challenges that this approach entails are therefore significant and it is high time to face them head-on. […] The advice is to be vigilant in terms of the technology chosen for connected objects as not all forms are worth it and "guzzle" electrical energy and data communications to varying extents."[8]

This was also stated by the director of the IEA, Maria van der Hoeven:

> "The proliferation of connected devices brings the world lots of advantages, but the cost is currently higher than it should be. Consumers lose money in the form of wasted energy, which leads to the establishment of more energy sources and structures than would normally be necessary, not to mention the increased greenhouse gas emissions."[9]

7 https://www.generation-nt.com/iea-rapport-appareil-electronique-internet-gaspillage-energie-actualite-1899762.html.

8 https://www.latribune.fr/opinions/tribunes/objets-connectes-de-l-eldorado-economique-au-fiasco-environnemental-491466. Specializing in industrial connected objects using RFID Active and Bluetooth, ELA Innovation is an important actor in the European market for identification using UHF (ultra-high frequency) radio waves.

9 https://www.generation-nt.com/iea-rapport-appareil-electronique-internet-gaspillage-energie-actualite-1899762.html. The IEA is an international body that provides analysis and data on energy, and policy recommendations and concrete solutions to help countries to provide safe and sustainable energy for everyone.

13.3. Digital technology: a powerful tool

Influencing consumers and guiding their behavior to improve their well-being or that of society are the main objectives of sustainable food-related communications. Such communication can occur as part of social marketing campaigns run by public organizations or with the aim of promoting corporate social responsibility. Digital tools, and especially mobile apps, are used to this end. Following a theoretical presentation of the different models, the observation of managerial practices will make it possible to draw up a typology of the digital devices that assist in the adoption of sustainable food.

13.3.1. *Successfully bringing about more responsible behaviors*

Five categories of models (Figure 13.1) allow us to understand how to influence consumers to bring about more responsible behaviors (Brennan *et al.* 2014): cognitive, conative, emotional, sociocultural and situational models. These models are not distinguished by the concepts used, but rather by what they highlight to change individual behavior: information, behaviors and habits, emotions, social influences or consumption.

Cognitive models share the idea that "humans behave in a consistent and rational manner" and that a decision is believed to be "logically planned and implemented, with clear decision-making criteria that allow the consumer to make the best possible decision for them, by weighing up the pros and cons of each decision and by choosing the option that is the best for them at a given moment" (Brennan *et al.* 2014, p. 17). In this regard, making additional information available strengthens the rational behavior of individuals. They can thus consider the costs and the benefits of a given behavior and decide whether to implement it. Three main levers have been identified in the literature on helping to bring about rational decision-making:

– perceived behavioral control: showing the individual that they are capable of implementing the behavior;

– subjective standards: showing them that an important person or group will approve and support the implementation of the behavior;

– the motivation and capability to process information: giving them the desire to process and information and helping them to analyze that information.

Conative models are focused on consumer behavior, paying less attention to feelings and thoughts. They share the idea that consumers "do not really make decisions, but take actions related to their past experience and the physical and social environment in which they find themselves" (Brennan *et al.* 2014, p. 41). They question the reasoning of individuals and emphasize the role of habits and routines in behaviors. Changing the behavior of individuals includes changing their social and physical environment or taking their decision-making biases into account[10]. Beyond the means aiming to reward or facilitate "good actions", a number of levers have been identified in the literature to help to bring about these behaviors and make them routine:

– barriers to the implementation of a behavior: reduce the barriers to implementing a behavior by making them simpler;

– feeling of personal capability: emphasize the perceived capability by implementing a behavior that will change things – reduce pollution – by using vicarious experience, coaching or providing information to show individuals that they are capable of implementing a given behavior;

– the advantage of the desired behavior in relation to the present behavior and its compatibility with the lifestyle of the individual: provide the consumer with proof that the desired behavior is beneficial to them and is consistent with their lifestyle;

– these models also emphasize the role of infrastructures in implementing a behavior.

Emotional models play on the emotions of the consumer (fear, blame, shame, etc.) to bring about behavioral change. People will adopt behaviors that will maximize their ratio of positive emotions to negative emotions. These models are particularly interested in what a consumer does in a situation of risk or uncertainty. Beyond the use of emotions such as fear, sadness, shame or anger, a number of levers are highlighted in the literature and are found in some of the other approaches mentioned:

– barriers to implementing behaviors;

– the negative consequences of the seriousness of current behaviors for individuals and those dear to them: show how the environmental or social damage can affect people and to what extent;

– the capacity of individuals to implement the behavior.

10 The theory on decision-making behavior shows that individuals often make decisions far from the optimum assumed by economic theory owing to, for example, a poor or partial perception.

Among these models, the theory of self-determination, in particular, shows that individuals have extrinsic and intrinsic motivations, and that the latter are more effective in changing attitudes and/or behaviors.

Sociocultural models account for the fact that people are social beings within a sociocultural system. More specifically, they consider sociocultural and sociodemographic differences and the relationships individuals have with others (family, friends, groups) in order to understand how to establish a behavior. This consideration can occur at either the personal level of relationships with parents, children or a spouse, at the meso-environmental level (neighborhoods, for example) or at the macro-environmental level (ideology, rituals). This involves playing on the individual's social identity as a lever for change by demonstrating that implementing the behavior would not be inconsistent with their social identity, or by emphasizing that the standards of those close to the consumer would support the implementation of the behavior. For example, starting to sort waste would allow the individual to be perceived as someone who is concerned about the environment. Being a member of an online community also makes it possible to play on socialization and reciprocal education among individuals.

Lastly, *situational models relating to consumer behavior* highlight the manner in which the individual makes a choice or makes a decision by identifying the decision-making process, the perceptual process and the environmental and individual factors. These models underline the importance of situations related to consumption, communication and shopping in determining behavior. Five components of the objective purchasing situation have been identified:

– physical environment: space, light, temperature, noise;

– social environment: the presence or absence of others;

– temporal perspective: season, time since last purchase;

– previous states: mood, anxiety, stimulation, illness;

– time pressure, i.e. the time that the consumer has to make the purchase.

Moving away from the situation, Jacoby (1978) and then Holbrook (1987) noted that consumption includes three practices: acquisition, use and disposal. *Acquisition* involves "the modalities of acquiring the product", *use* corresponds to "the different modes of functional and ergonomic interactions with objects" and *disposal* concerns "all practices involved in disposing of an object (rejection, transformation, recycling, etc.)". Behavioral change must thus include a deep understanding of all of these practices.

Cognitive models	Provide information	Improve the cost/benefit determination in favor of implementing behavior
Conative models	Change the environments of individuals and take decision-making biases into account	Change routine and habitual behaviors
Emotional models	Use emotions, such as fear, sadness or anger	Maximize the ratio of positive to negative emotions
Sociocultural models	Awareness of the sociocultural system which the individual is a part of	Play on the individual's social identity and social influence
Consumer behavior models	Take into account the consumption context and the decision-making process of the consumer	Provide the consumer with the tools to modify the way in which they buy, use and dispose of products

Figure 13.1. *The five models for influencing the consumer to create more responsible behaviors*

13.3.2. *A typology of digital tools according to their objectives*

Two axes enable us to place all digital devices in a structure (Figure 13.2): benefits provided (utilitarian versus hedonic) and temporality (purchase versus consumption).

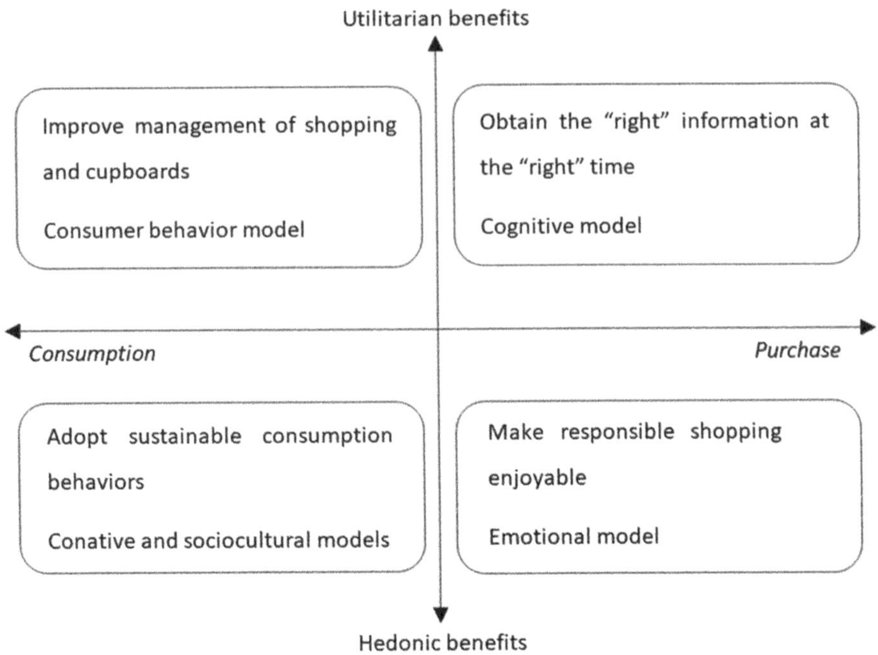

Figure 13.2. *Overview of the role of digital devices in assisting responsible food consumption*

13.3.2.1. *Objective 1: improve management of shopping and cupboards*

Making useful information available to consumers in their homes is an objective of the digital tools that invade their daily lives and kitchens. This concerns all domestic appliances and kitchen equipment. For example, the website super-marmite.com allows consumers to gather recipes, tests, comparisons and tips.

As part of this simplification of day-to-day management, Monoprix also offers the option of creating a shopping list with its voice assistant[11]. Consumers can ask to "talk to Monoprix" and add the product they want after having set up their loyalty card.

A lot of apps and connected packaging are also intended to help the consumer in managing their stores and reducing household food waste. To that end, IBM

11 https://www.actionco.fr/Thematique/solutions-motivation-1261/breve/le-phygital-revolutionne-le-retail-et-les-programmes-relationnels-336286.htm.

developed "Watson", which, thanks to artificial intelligence, can combine ingredients to suggest recipes, while taking into account the specific expectations of consumers (allergies or foods they do not eat).

13.3.2.2. *Objective 2: obtain the "right" information at the "right" time*

Many digital apps have emerged in the food domain. Given the ubiquity of the smartphone, these digital tools support consumers in their movements and their decision-making processes. They provide the "right" information at the "right" time (see Box 13.1). They offer additional information: labels from which the product benefits, the nature or composition of the product, and the availability and properties of sales outlets. For example, the Planet Ocean[12] mobile app aims to protect maritime resources by making information available on endangered areas and sustainably managed resources (fishing techniques, fish stocks, etc.) and guiding consumers towards responsible purchases.

This information is intended to be used in decision-making processes, for example when making a purchase in store. As a result, the presentation must be concise. The Noteo app, for instance, allows users to discover a product's rating by scanning it. This rating, which has a scientific purpose, is based on "the guidance of experts in nutrition, the environment, toxicology and CSR"[13].

Such ease of use is also reflected in the potential entries. The Flash Labels app, providing information about the characteristics and trustworthiness of environmental labels, allows users to search for a key word and using an "innovative image recognition function using the photo app of an iPhone"[14].

Furthermore, the information can refer to sales outlets thanks to geolocation technology. The Etiquettable[15] app, for instance, provides a list of retailers who can supply locally produced fruit at a local level. Similarly, Green Raid[16] provides a collaborative platform that offers consumers the chance to identify places in France for local and responsible consumption.

---

12 https://www.goodplanet.org/fr/planete-ocean-1-film-pour-comprendre-1-appli-pour-agir/.

13 https://www.consoglobe.com/noteo-lappli-qui-unifie-la-notation-des-produits-de-consommation-cg.

14 https://www.entrepreneursdavenir.com/actualites/flash-labels-la-premiere-application-mobile-sur-les-labels-environnementaux/.

15 https://etiquettable.eco2initiative.com/.

16 https://www.greenraid.fr/.

13.3.2.3. *Objective 3: adopt sustainable consumption behaviors*

Sustainability involves abandoning habits and routines and the repetition of behaviors. The restrictions affecting these new behaviors can explain, at least partially, the difficulty of adoption. In this regard, digitalization can reduce the sacrifices made and/or increase the benefits enjoyed, in particular by adding emotional or entertainment value. For example, the game "Reductor"[17], available on smartphones, is a communications tool implemented jointly by ADEME (French Environment and Energy Management Agency) and the French Ministry for Sustainable Development. It brings entertainment together with raising awareness of how to reduce waste.

While recipes were traditionally passed on from generation to generation and reflected private practices, they are now "digital" and are conveyed via YouTube channels, Instagram accounts and numerous blogs dedicated to the subject. Sharing, a key part of cooking, takes place through sharing communities on social networks. These communities forge strong links by sharing recipes and exchanging tips and guidance, among other things.

This assistance in everyday cooking tasks partially explains the success of Thermomix, as indicated on its website: "Your personalized Cookidoo® fits perfectly on the fantastic screen of your new Thermomix®, giving you direct access to a world of inspiration, with over 45,000 recipes from all over the world and ideas to make simple, homemade cooking a success."[18]

13.3.2.4. *Objective 4: make responsible shopping enjoyable*

Digitalization has caused significant changes in food purchasing, as shown by the experience of Nespresso: "Digital technology makes it possible to give a leading role to the sales experience and to the sellers in particular. The latter, equipped with tablets, are genuine coffee specialists, able to teach customers about coffee tasting rituals or provide them with details on Nespresso manufacturing processes."[19] Hedonic benefits are also enhanced by digital devices. For example, the app Too Good To Go offers a surprise box made up of products approaching their use-by dates.

---

17 https://presse.ademe.fr/2010/12/lancement-du-jeu-reductor-stop-aux-dechets-sur-iphone-1.html.

18 https://www.vorwerk.fr/thermomix/page-daccueil/.

19 https://www.actionco.fr/Thematique/solutions-motivation-1261/breve/le-phygital-revolutionne-le-retail-et-les-programmes-relationnels-336286.htm#7I6JeGeVPvukQ0k6.97.

Launched on April 2, 2019, the app MyLabel aims to provide concerned consumers with trustworthy and personalized information to buy more responsible products. Two specific features distinguish MyLabel from other apps on the market:

– consumers must choose the criteria (environmental, social or health) that they want to be evaluated;

– the information available comes from consumer associations or public bodies, sources of information considered credible by consumers.

While everyone agrees on the need for change, companies and consumers find obvious reasons not to engage with it. It has to be said, "for a company, change is risky. You can always lose your customers because they don't like changing their habits. They say 'I'm done with Nutella', but then when there's a promotion they come running". This statement points to the heart of the MyLabel mobile app: "And so, we make it possible for them [companies] to reduce the risk of such change. Your customers want to go in this direction and they are prepared to go there. They start to buy. So, let's go, it's now, the risk is low and you aren't going to lose money." That was all it took to start the virtuous circle.

For Christophe Hurbin, four main factors make it possible for such entrepreneurial projects to emerge. Firstly, consumers have changed: "For a while now, for between ten and fifteen years, European and French consumers, in big surveys, have been telling you that, for half of them, they want products that are more environmentally friendly and they are prepared to buy them. They don't always necessarily try to do so. The intention is there but, on the other hand, if you help your customer to make good on their intention, you are much more powerful than if, despite their intention, you continue to sell them something they buy despite themselves or that they buy out of habit." Beyond this intention, consumers are better informed and better educated. "This education means that with things they don't know, when they find them out they become shocking, uncomfortable, even against our principles. We would like to make them stop, or we would, in any case, not like to be an actor or complicit in some things."

At the same time, technical progress and technological developments have made it possible to make information accessible and reliable. Christopher Hurbin noted that: "The second thing is that information is available and that, really, is about technological means. Because the information that palm oil cultivation kills orangutans has been available on the Internet for a long time. On the other hand, connecting that with the product I'm about to buy is really recent."

Thirdly, the culture of companies is changing and highlighting this willingness to support responsible consumption. Companies and their employees "want to do something even if there are constraints on the business".

Lastly, the pioneering app Open Food Facts plays a key role in making information available: "I think that Open Food Facts has really been a trigger, (1) it has made free apps possible, on a cost scale, apps that have the time to find their business model, etc. but (2) it has also put pressure on companies that now do not want their products to be assessed on the basis of incorrect information."

In this regard, the MyLabel app occupies an intermediary position and is differentiated by an original offer for consumers and companies. On the one hand, the app offers consumers a personalized approach based on multiple criteria, and where information is collected by a real body in the field of consumption. Consumers can thus choose their criteria and give meaning to their consumption: "I also really believe in an approach that doesn't look to blame people. I think it's really essential that, if we want people to change, we mustn't blame them, they have to feel motivated, supported." On the other hand, the app is also interested in companies: "This is an axis that is 100% marketing when we suggest that they integrate what users are saying, that is their attitudes, their intentions, their general profile, anonymously of course, in order to find out about consumer trends. Is animal welfare becoming more important? Are people buying organic products more aware of such things? Stuff like that. And this increases with real purchases to see whether people are moving from intention to action." The use of this data is central to the app's business model: "We want this virtuous circle where, at the end, we make companies change. Because if you go through consumer surveys, they tell you that the actors they trust are associations but that the actors that will change things are companies because they have the means. Taking information from associations to make companies change makes a lot of sense in relation to consumers."

Box 13.1. *Interview with Christophe Hurbin, Co-founder of MyLabel*

13.4. Conclusion

Although it highlights the contributions of digitalization to the gradual shift towards a more responsible consumption, this chapter invites companies to reflect on the technological characteristics of the tools used, as well as to consider the potential obstacles related to their intrusiveness.

Consumers, in fact, display a lack of trust in the commitments of companies to socially responsible consumption, for example, commitments relating to food waste. Some consumers doubt the sincerity of the motivations of organizations by comparing those who take on ethical commitments or establish pro-environmental actions for selfish reasons (to boost their image) and those who do so for altruistic reasons (Karaosmanoglu *et al.* 2016), with this suspicion being greater for commercial companies than for associations (Terwel *et al.* 2009). To reduce this obstacle, MyLabel suggests relying on cooperation with associations that have guarantees of the credibility of their information. Proving the legitimacy of organizations to offer people a meaningful mobile app is of the utmost importance to reduce intrusiveness. The historical legitimacy of the means and objectives allows us to identify three areas.

Historical legitimacy: highlight the consistency between the means and the actions of the company; demonstrate skills

Importance of storytelling: BuyorNot presents the history of its app on its website: "Founded in 2015 by brothers Bulent and Levent Acar from Lyon, the association i-buycott aims to raise public awareness of responsible consumption. Its main tool is the collaborative platform i-boycott.org, launched in 2016. With over 100,000 active users, i-boycott is the tool that offers people the opportunity to express their demands of economic actors via boycotting campaigns. Oasis, Philips, Petit Navire, etc. A number of large groups have changed their practices because of the mobilization of boycotters! So, ready to join us?"[20]

Legitimacy of means: indicate the skills and knowledge of the company in relation to the relevant issue

It is possible to present the original team in detail, as well as the committee of experts around the app, as in the case of Etiquettable, in order to reassure the consumer of the company's knowledge[21].

Legitimacy of objectives: highlight the company's altruistic motivations and potentially partner with associations to reassure consumers

The BuyOrNot website indicates that: "BuyOrNot is a mobile app launched by i-buycott, a public interest association under the 1901 Act. The app helps consumers to make ethical purchases by allowing them to scan their chosen products in store. A practical app, BuyOrNot demystifies the labels on every item by revealing their impact on health and society. Thanks to this app, it's impossible to make bad choices!"[22]

Box 13.2. *Areas of action with a view to proving the legitimacy of the organization to offer a meaningful app*

The possibility of personalizing tools seems to be an area of interest in limiting invasive intrusiveness. In the case of mobile apps, this can involve personalizing the ergonomic design, as well as the notifications. In order to reduce interfering

20 https://buyornot.org/qui-sommes-nous/.

21 https://etiquettable.eco2initiative.com/experts/.

22 https://buyornot.org/qui-sommes-nous/.

intrusiveness, MyLabel suggests to consumers themselves that they determine their chosen criteria to improve their consumption in the priority areas[23]. This allows consumers to feel less infantilized. It is then necessary to reassure consumers in relation to the use of data and presence of advertising.

Lastly, the designers must take a number of variables into account, such as ease of use, interest, enjoyment, informative nature and perceived savings, and potentially check these tools prior to their launch by testing the mobile app among individuals and by taking note of user comments on the platforms where apps can be downloaded online. For example, the following opinion of Yuka is available on Google Play: "Very good app. Recommended. Pity there aren't any recommended substitutes for all the products scanned. Being able to search by product would be great (example search: Brie; result: all products associated with the term would have a score of excellent or good, etc.)." The company responded: "Hello, it is possible to carry out searches and filter the results; this is a functionality in the premium version of the app. This paid version helps us to keep developing the project and maintain our independence." In the case of apps, such opinions are a very rich vein of information that can be mined to find ways of improving.

13.5. References

Brennan, L., Binney, W., Parker, L., Aleti, T., Nguyen, D. (eds) (2014). *Social Marketing and Behaviour Change: Models, Theory and Applications*. Edward Elgar Publishing, Cheltenham.

Comino, L. (2017). La numérimorphose des courses ordinaires : une approche par les usages. PhD thesis, University of Lorraine, Nancy.

Cova, B. and Cova, V. (2009). Les figures du nouveau consommateur : une genèse de la gouvernementalité du consommateur. *Recherche et applications en marketing*, 24(3), 81–100.

Fuentes, C. and Sörum, N. (2019). Agencing ethical consumers: Smartphone apps and the socio-material reconfiguration of everyday life. *Consumption Markets & Culture*, 22(2), 131–156.

Gholami, R., Sulaiman, A.B., Ramayah, T., Molla, A. (2013). Senior managers' perception on green information systems (IS) adoption and environmental performance: Results from a field survey. *Information & Management*, 50(7), 431–438.

Giesler, M. and Veresiu, E. (2014). Creating the responsible consumer: Moralistic governance regimes and consumer subjectivity. *Journal of Consumer Research*, 41(3), 840–857.

Gonzalez, C. and Siadou-Martin, B. (2019). Vers une clarification de l'intrusivité des applications mobiles servicielles : le cas du gaspillage alimentaire. *Décisions Marketing*, 94, 13–34.

23 https://www.mylabel.io/#download.

Gonzalez, C., Siadou-Martin, B., Ferrandi, J.M. (2019). Proposition of a scale of servitized mobile apps' perceived intrusiveness: The case of the fight against food waste. *6th French-Austrian-German Workshop*. Nancy, 20–22 November.

Karaosmanoglu, E., Altinigne, N., Isiksal, D.G. (2016). CSR motivation and customer extra-role behavior: Moderation of ethical corporate identity. *Journal of Business Research*, 69(10), 4161–4167.

Lecompte, A.F. and Valette-Florence, P. (2006). Mieux connaitre le consommateur socialement responsable. *Décisions Marketing*, 67–79.

Lovbrand, E. and Stripple, J. (2017). *Bringing Governmentality to the Study of Global Climate Governance, in Governing the Climate: New Approaches to Rationality, Power and Politics*, Stripple, J., Bulkeley, H. (eds.). Cambridge University Press, Cambridge.

Miller, P. and Rose, N. (2008). *Governing the Present: Administering Economic, Social and Personal Life*. Polity Press, Cambridge.

Murugesan, S. (2008). Harnessing green IT: Principles and practices. *IT Professional*, 10(1), 24–33.

Rodhain, A. (2013). "J'aimerais bien, mais j'peux point..." : exploration des écarts entre attitude et comportement en consommation alternative. *Management & Avenir*, 3, 50–69.

Séré De Lanauze, G. and Siadou-Martin, B. (2016). Dissonance cognitive et stratégies du consommateur face à une information critique : le cas de la consommation de viande. *Revue française du marketing*, 257, 39–60.

Sirieix, L. (1999). La consommation alimentaire : problématiques, approches et voies de recherche. *Recherche et applications en marketing*, 14(3), 41–58.

Stripple, J. and Bulkeley, H. (2017). Introduction on governmentality and climate change. In *Governing the Climate: New Approaches to Rationality, Power and Politics*, Stripple, J., Bulkeley, H. (eds.). Cambridge University Press, Cambridge.

Terwel, B.W., Harinck, F., Ellemers, N., Daamen, D.D. (2009). How organizational motives and communications affect public trust in organizations: The case of carbon dioxide capture and storage. *Journal of Environmental Psychology*, 29(2), 290–299.

Uggla, Y. and Soneryd, L. (2016). Green governmentality, responsibilization, and resistance: International engos' issue framings of future energy supply and climate change mitigation. *Socijalna Ekologija*, 26(3), 87–104.

14

Augmented Products: The Contribution of Industry 4.0 to Sustainable Consumption

Myriam ERTZ, Shouheng SUN, Émilie BOILY, Gautier Georges Yao QUENUM, Kubiat PATRICK, Yassine LAGHRIB, Damien HALLEGATTE, Julien BOUSQUET and Imen LATROUS

LaboNFC, University of Quebec at Chicoutimi, Saguenay, Canada

14.1. Introduction

Planned obsolescence has often been decried by researchers, politicians, the media and the general public, but what is it? Technically speaking, obsolescence is the depreciation of materials or equipment ahead of their material wear. Planned obsolescence, for its part, is a strategy used by manufacturers aiming to reduce the life spans of objects via various means: by making them unrepairable; by reducing the guarantee; by making replacement parts unavailable or expensive; by using low-quality materials; or by shortening some assembly stages. This, at any rate, is what was suggested by Vance Packard, an American theoretician working on planned obsolescence in the 1970s. However, is this obsolescence planned? That is the real question, as very often it is a question of lower costs to maintain favorable prices and stay competitive. In this case, obsolescence would therefore be involuntary. Nonetheless, whether it is planned or involuntary, the results are the same: significant waste production, especially electronic waste. In 2016, 44.7 million tons of waste formed of electronic devices was generated worldwide. By 2025, this volume is predicted to rise to 63.7 million tons (Research and Markets 2018).

Marketing for Sustainable Development,
coordinated by Sihem DEKHILI.

The company Apple has often been criticized for having used a widespread strategy of planned obsolescence to increase its sales, and the group has been the subject of numerous complaints in that regard (Rozières 2017). However, Apple is not the only company targeted. In 2017, Greenpeace and the repair site iFixit launched the #RETHINKIT initiative in which they identified 17 brands carrying out planned obsolescence. This list included Apple (Retina Macbook, Macbook Pro, iPad) at the top, as well as Samsung (Galaxy S7, S7 Edge and S8) and Microsoft (Microsoft Surface Pro 5, Surface Book) (Rozières 2017). Robin Perkins of Greenpeace even confirmed that "this is planned obsolescence. The trend is to design products that are ever less repairable and ever more disposable in order to boost sales" (Rozières 2017). Faced with this deluge of criticism, legislators decided to intervene. For example, in 2019, the French National Assembly voted almost unanimously for a repairability index in 2021. This concerns electrical appliances, a measure included in article 2 of the circular economy bill. This legislative intensification can be perceived as a restriction, but it is also an opportunity for companies to make their production processes and their products more responsible. In order to achieve this objective, technological solutions are currently being developed.

Industry 4.0, in general, and Big Data, in particular, are promising from the perspective of more sustainable development. Technology and sustainable development can be seen as "titans [that] clash" (Jabbour *et al.* 2018). The metaphor is an eloquent one and is based on the premise that sustainable decision-making processes require very precise information and data, and information technologies (IT) are increasingly able to provide it. According to this premise, a better understanding of the environmental impact of products, of their movement through the supply chain, of their use by customers and of what happens to them at the end of their lives will provide crucial data. Indeed, they could be used by organizations to make products more sustainable. Moreover, such data comes increasingly often from so-called "smart", "augmented", "connected" or "autonomous" products. Their online interconnection creates the famous Internet of Things, able to generate artificial intelligence. These data can thus be used in production processes and, in particular, in additive manufacturing (3D printing). Here we can see the emergence of a promising technological quartet: Big Data; the Internet of Things; artificial intelligence; and additive manufacturing.

In marketing, the use of such rich and varied data would provide a complete picture of the life span of a product. With connected objects, we do not only capture a 360-degree image of consumers and their behaviors during purchases and at given moments, which would typically only occur through data mining[1]. These technologies also permit a thorough understanding of the pre-purchase, purchase, use and post-use stages of the product by the consumer in real-time. The behaviors of stakeholders related to the product could also be recorded. Big Data would thus contribute to a refined understanding of the life span of products. The design, production, maintenance, redistribution, repair and reclamation stages of the product's life could also be analyzed. Such progress in sustainability, rooted in technology, could contribute to creating a new way of marketing products. This would be supported by the extension of the life span of the product to create further value and potentially combat planned obsolescence.

The concept of planned obsolescence, or rather the perception of it, is closely linked to the life span of the product. This is defined as "the duration of the period beginning at the time a product is put into circulation and used that then ends at the time when a product becomes obsolete without the possibility of repairing the product" (den Hollander *et al.* 2017, p. 519). The extension of the life span of the product refers to the means used to extend this period. This excludes end-of-life processes, such as recycling (Den Hollander *et al.* 2017), destruction for reuse, composting or biodegradation. The life span of the product is extended by means of a variety of bodies generically referred to as "business models for extending the life spans of products". These are intended to improve product design (nature-based strategies) or increase the life span of the product during the use and post-use stages (nurture-based strategies)[2] (Ertz *et al.* 2019).

Nevertheless, despite the perceptible links between technology and sustainability, few studies examine the opportunities for crossover between the two fields (Jabbour *et al.* 2018). As a result, the potential synergies between technological innovation and sustainability to improve the life spans of products are little known. This chapter thus answers the following question: to what extent can IT (commonly known as Industry 4.0) contribute to extending the life spans of products in order to create a new marketing of "augmented" products in support of sustainable consumption? Industry 4.0 has two heavily interrelated components.

1 Data mining aims to analyze large quantities of data from different sources (for example, management software packages, customer relations management software) using algorithms, in order to extract useful knowledge or know-how from it for managerial decision-making purposes.

2 A nature-based strategy is aimed at improving the design of a product. A cultivation strategy, for its part, is intended to preserve the product throughout its life span using repair methods, for example.

On the one hand, there are infrastructures and processes, whether physical systems such as 3D printers or connected objects that generate and use data. On the other hand, Industry 4.0 has analytical capacities, whether Big Data or artificial intelligence, that are primarily based on the same statistical methods and algorithms. These enable the use of data generated by infrastructures and processes and transform them into business intelligence. Lastly, these data, when analyzed, can be used to support decisions made during these processes.

Under the theoretical framework of business models for extending the life spans of products (Ertz *et al.* 2019), there are five activities for extending the life span of a product, namely:

1) *Improvement of product design*: activities to improve products and production processes, as well as improvements to opportunities for maintenance and repair.

2) *Access*: services focused on product use, including the rental, loan, sharing and pooling of products.

3) *Maintenance*: services focused on product care through agreements on maintenance, advice, training and consultancy.

4) *Redistribution*: activities for transferring products from one individual to another through various exchange-related activities, such as donations, swaps, and second-hand purchases and sales.

5) *Reclamation*: restoration of a product to its original state of operations or an updated state through repairs, refurbishment, remanufacturing or reprocessing of the product.

Box 14.1. *Theoretical framework of business models for extending the life spans of products*

This chapter presents a crossover between the literature on Industry 4.0 and on extending the life spans of products (sustainable consumption). We will use the theoretical framework of business models for extending the life spans of products (Ertz *et al.* 2019), presented in Box 14.1. This will enable us to identify IT's ability to contribute to prolonging the life span of products.

14.2. Infrastructures and processes

14.2.1. *Additive manufacturing and shifts in production paradigms*

Additive manufacturing (AM), or 3D printing, is a range of production techniques that facilitate more flexible manufacturing modes (Sauerwein *et al.* 2019). Numerous recent studies have explored the potential of this technology for sustainability (for example (Van Wijk and Van Wijk 2015; Sauerwein *et al.* 2019)). For example, an innovative strategy involving AM enables parts to be manufactured directly from parts at the end-of-life (EoL) stage (or existing parts). This avoids the material recycling stage, which can extend the life span of the product. AM shifts the design paradigm and offers greater opportunities for product innovation. AM has six main characteristics: high degree of freedom in design, great complexity of form, personalization and adaptation, localization, sustainability, standardization and compatibility.

AM offers a *high degree of freedom in design*. Products can have several lives thanks to upgraded pieces, by replacing old parts, and/or through redesigning or restyling[3]. For example, clothing can be adjusted to accommodate personal needs, such as in sizes, shapes, preferences and so on. AM would thus enable a significant extension of the life span of a product.

Moreover, AM enables a *great complexity of form*. As a result, digital production has to take control of repairs as broken parts can be imitated and reproduced. AM is thus recognized as a production technique that can encourage repairs. Printing separate parts could also make it possible to extend the life span of a product by improving the possibility of refurbishing and repairing it. This technology can help recover components with specific weaknesses by adding materials or by replacing those discarded by producing new ones. For example, some laser fusion technology can be used to reconstruct more complex parts, repair broken parts and improve manufacturing processes. Moreover, products can be designed to facilitate maintenance and repairs. This can keep the product in good condition by enabling repairs and the replacement of broken parts. At the same time, products can also be better designed in terms of scalability and adaptability by incorporating extendible and modifiable options. They thus continue to be useful in changing conditions and keep their quality, value, efficacy and performance (Sauerwein *et al.* 2019).

---

3 Restyling is aimed at modernizing or at least updating the style of a product in order to make it more contemporary, which can increase psychological acceptance of the product.

AM enables *personalization and adaptation* in product design. While traditional technology leads to changes in production cycles, molds, tools or assembly lines to personalize products, this is not necessary with AM. It is enough simply to digitally modify the models to be produced. 3D scanners and modeling tools thereby seamlessly contribute to personalization in production or mass personalization (Sauerwein *et al.* 2019). Personalization and adaptation are considered design strategies for creating a stronger user–product relationship. This personalized and adapted design could, to a certain extent, lead to longer life spans for products.

Another important feature of AM is the localization of production, maintenance and repairs. Localized production shortens the supply chain and therefore requires less transportation. This reduces the deterioration of the product and extends its life span. The broken parts of an (old) product can be printed *in situ*, which solves the problem immediately. The life span of the product can thereby be extended more effectively (Van Wijk and Van Wijk 2015). AM would be able to improve the *sustainability* of a product by designing products individually in an artisanal manner. This thus suggests a greater degree of freedom in increasing the reliability of using more sustainable materials and components, as well as by making specific arrangements, contours, folds or cuts (Sauerwein *et al.* 2019). The features of localization and personalization/adaptation in AM can facilitate the co-design and/or co-creation of a product by a family, friends or a community (Van Wijk and Van Wijk 2015). The life span of products can be extended by sharing and pooling between people or groups. Moreover, the co-created products often have a higher value, which facilitates their redistribution.

AM also enables the *standardization and compatibility* of products. It makes it possible to create products with parts or interfaces adapted to other products (products of the same or similar type). This improves the redistribution of products through second-hand transactions and exchanges. Moreover, this facilitates the emergence of new service models focused on usage, such as the renting, sharing or pooling of products (for example, a cell phone charger).

In accordance with the above, Table 14.1 provides a clearer picture of the potential ways in which AM can contribute to extending the life span of products.

Key activities	Features and functionalities of AM						Contributions to extending the life spans of products
	High degree of freedom in design	Great complexity of form	Personalization and adaptation	Localization	Sustainability	Standardization and compatibility	
Improvement of product design	✓	✓	✓		✓		– Sustainable materials and design to improve the reliability and sustainability of the product – Design a product in a way that facilitates maintenance and repairs – Scalable and adaptable product design, incorporating extendible and modifiable options, to remain useful in changing conditions and to improve quality, value, efficacy and performance – Personalized design adapted to the product
Access	✓	✓	✓	✓		✓	– The localization and personalization features of AM can lead to improved sharing and pooling of products between a number of people or groups – Standardized and compatible products facilitate the promotion of new service models based on use
Maintenance	✓	✓		✓		✓	– Localized production can shorten the supply chain and reduce transportation, thereby reducing the deterioration of the product and extending its life span – The broken parts of any (old) product can be printed *in situ*, which resolves the problem immediately

Key activities	Features and functionalities of AM						Contributions to extending the life spans of products
Redistribution	✓	✓	✓	✓		✓	– Products are standardized and compatible with parts or interfaces that can be adapted to other products (products of the same or similar types), facilitating redistribution through second-hand transactions and exchanges – The products designed are of a greater value, have a longer life span and are thus easier to redistribute
Reclamation	✓	✓		✓	✓	✓	Contribution to the repair or refurbishment of products to give them "multiple lives" by improving parts, replacing old parts, and/or rethinking or remodeling parts and components. For example, clothes can be adjusted to accommodate individual needs, such as sizes, shapes and preferences. Furniture can be adapted to personal circumstances, shapes and preferences, and special elements can be added to products at a later date

Table 14.1. *Features of the contribution of additive manufacturing to extending the life spans of products*

14.2.2. *The Internet of Things in favor of the automated and remote management of products*

The Internet of Things (IoT) refers to connected devices, typically with sensors or RFID (radio-frequency identification) chips. It enables objects to communicate and become active participants in an information network such as the Internet (Joshi and Gupta 2019). The IoT transforms autonomous products, such as a table, refrigerator or microwave, into smart products that are connected (to the Internet).

Thanks to the IoT, companies can thus carry out monitoring, from a distance, during use and in real-time, of the condition and location of products. The condition of products takes into account the level of wear, functioning, use and current environment of the product. Identification and localization technologies enable the products themselves to be monitored at the level of their components. The IoT makes data available via an infrastructure network. The IoT and its features can be used to develop feedback-rich systems and loops through the life cycle of a product. Through the adoption of the IoT and its corollary data collection during product use, companies can optimize the use of goods by using the information generated throughout the life cycle of the product. This allows them to optimize maintenance, reuse, repair and refurbishment activities.

The IoT has four major features: surveillance and monitoring of product activity; remote control and self-monitoring; the creation of databases on the life spans of products; and support for analytics and business intelligence.

First of all, the IoT enables manufacturers to carry out *surveillance and monitoring of product activity*. Manufacturers can obtain information on how users interact with the product and learn more about the habits and behaviors involved in the use of the product. This information can then be used to improve product design (Porter and Heppelmann 2014). On the one hand, manufacturers can optimize product design in line with the real use of the product. On the other hand, they can design products that are easier to maintain, upgrade, disassemble and recycle.

For example, in the Advanced-Remanufacturing-To-Order-Disassembly-To-Order system, sensors and RFID chips are integrated into objects in order to monitor key components during the economic life of the product. They then deliver all of the information collected during the life cycle of the product when it reaches the reuse or recycling stage (Joshi and Gupta 2019). More comprehensive knowledge of the conditions and quality of EoL products, as well as an estimate of their remaining life span, makes it possible to do away with an expensive preliminary disassembly and time-consuming inspections. This leads to optimal planning and remanufacturing (Joshi and Gupta 2019).

Secondly, the IoT includes *remote control and self-monitoring*. Companies control the functionalities of the product and personalize the user experience from a distance. The corollary of remote control is self-control or autonomy, in the sense that the product is capable of carrying out self-diagnostics, self-coordination or even autonomous functioning. Carrying out tasks from a distance or autonomously has profound implications for the marketing of services and for relationship marketing. It makes it possible to improve the efficacy and security of service-delivery processes, in particular for companies offering remote services (such as call centers, telesales services), monitoring, telecommunications, IT services, repairs or maintenance, in which the client could be encouraged to (partially) contribute (Porter and Heppelmann 2014).

Thirdly, through the constant interaction of connected objects and devices, the IoT enables the creation of *databases on the life spans of products* by recording and continuously integrating information on the goings-on of a specific product, with this information supplemented by remote monitoring activities. The databases built in this way accumulate statistical information on the design, production, disassembly and reliability and information on the disposal of the product.

Moreover, dynamic information is collected (or the history of the product), such as the remaining useful life span, use, maintenance, replacement of parts or refurbishment history. The creation of databases is an important feature. The integration of different data types from different sources is essential to ensuring the successful management of the life spans of products.

Lastly, the vast quantities of information accumulated in databases throughout the life span of the product provide unprecedented information on the usage phase, the functioning of the product and user characteristics (Porter and Heppelmann 2014). This makes it possible to optimize *support activities for analytics and business intelligence*, with the ultimate aim of facilitating decision-making (Lee and Lee 2015).

Based on these four functionalities, the IoT can extend the life spans of products in several ways, as presented in Table 14.2.

Key activities	Features and functions of the IoT				Contributions to extending the life spans of products
	Surveillance and monitoring of product activity	Remote control and self-monitoring	Databases on the life spans of products	Support for analytics and business intelligence	
Improvement of product design	✓		✓	✓	– By collecting and analyzing the usage data on how customers use the products via the IoT and through suitable analyses, companies can improve the design of products for users – By using the databases generated by connected products through their life cycle, manufacturers can use this history and the related analytics to optimize product design strategies for the entire life cycle (for example, to optimize maintenance, upgrading, disassembly, reassembly and recycling)
Access	✓	✓	✓	✓	– Monitor the condition, location and use of the product; the related information can be collected and easily made available to each customer, thereby facilitating the sharing of products between several users – Promote the emergence of new models of using products and new commercial models (for example, the widespread collaborative consumption of all types of products)

Table 14.2. *Features of the Internet of Things and its contribution to extending the life spans of products*

14.3. Analytical capabilities

14.3.1. *Big Data: a 360-degree knowledge of the product*

Big Data plays a key role in the development of designing products, systems and services. It makes them more intelligent, better connected, more efficient and more accessible. Data collection during the production process enables each stage of the process to be monitored. These data can be used to model processes, which can then be optimized in turn. Modeling Big Data offers a significant advantage. The models created can be more precise than the models created using fewer data. A model based on Big Data combining data on product, material and process parameters can provide new product design and development solutions. It can also considerably reduce errors during manufacturing and remachining (Chan *et al.* 2010).

Big Data is characterized by the four Vs: *volume* (large amounts of data), *variety* (data from inside and outside the company, either structured in the form of databases or unstructured, such as social media content), *velocity* (data produced quickly) and *veracity* (data providing a reliable account of the features of products, materials and processes).

The possibilities that Big Data offers to extend the life spans of products are analyzed below, based on the theoretical framework provided in Box 14.1:

– *Improved design*: on the one hand, manufacturing data can directly contribute to establishing repackaging parameters (Marwah *et al.* 2011). On the other hand, user behavior data linked to the product can continuously provide the manufacturer with new information about the design (Joshi and Gupta 2019). The optimization of product design can thus improve the quality of products and accessories, thereby extending their life spans (Joshi and Gupta 2019). For example, the new demand trend mining (DTM) technique allows companies to optimize the design of their products by simultaneously considering the pre-life stage (manufacturing) and the end-of-life stage (refurbishment) of a product.

– *Maintenance and refurbishment*: Big Data can help manufacturers assess the general condition of a product, as well as the remaining life span of the product's components, in order to determine the length of the guarantee that should be offered to customers who buy components and refurbished products. Big Data plays a key role in facilitating the development of guarantee and maintenance procedures for refurbished products by considerably reducing or eliminating almost all ambiguity about the existence and condition of the remaining components at their end-of-life. A simulation based on Big Data allows one to better evaluate interventions in the life cycle of the product and to identify the optimal strategy and moment of intervention in the ongoing life of a product (for example, when to update, refurbish, remake or recycle it in order to extend the life span of the product) (Charnley *et al.* 2019). In this context, Ciora *et al.* (2015) propose improving the quality of a product by analyzing Big Data in order to find, in particular, the best solution for repairing faulty vehicles in line with their characteristics. Moreover, a recommendations system based on Big Data from real-time monitoring can provide recommendations to prevent or mitigate errors and provide information on the possible reasons that these parts have become defective (Ciora *et al.* 2015). Ding *et al.* (2018) developed a cost analysis model to determine the complicated relationship between data obtained from sensors and the corresponding reprocessing processes, as well as the associated costs. They then extract the Big Data and optimize the swarm of backpropagation (BP) particles to predict the refurbishment costs, which can also contribute to hasty decision-making in relation to the manufacturability of products when they reach their end-of-life. Based on significant data on the life cycle of products, Fang *et al.* (2016) used a particle swarm optimization (PSO) algorithm to optimize the recovery strategy linked to the refurbishment, reuse and disposal of components in a PC society. Lastly, we note that Daimler Trucks North America improved its maintenance program for commercial vehicles by automating and rationalizing the refurbishment process. The manufacturer used Big Data and blockchain technology with the aim of tracing and verifying the origins and history of the parts, and thus built trust in data thanks to its improved security (Informa Commercial Vehicle Staff 2018).

Access and redistribution: the analysis of Big Data of consumer products allows us to better understand consumer behavior and their interventions in the product use process. This enables the promotion of new product-service use and redistribution models. For example, the system for evaluating the quality of Big Data can help the company make decisions on the manufacturing of new products and on the refurbishment of products. This can also facilitate the rebuilding of products returned after years, enabling the manufacturer to sell the two products simultaneously. Smart products intertwined with the Internet of Things will collect and communicate data on user behavior and the condition of the product. These data could be used to facilitate the management of systems for rentals, free or fee-based loans, sharing or pooling (access-based consumption). Moreover, owing to the transparency and free flow of information on users and products that Big Data provides and with the support of other technologies, products can be better redistributed through exchanges such as swaps, donations and second-hand sales (redistribution).

Key activities	Features of Big Data				Contribution to extending the life spans of products
	High volume	Variety	Velocity	Veracity	
Improvement of product design	✓	✓	✓	✓	– Manufacturing data can provide information on the redesigned parameters for optimizing product design to improve the quality of products and thus extend their life span – Data on product-related user behavior continuously provide the manufacturer with new information useful to product design
Access	✓	✓	✓	✓	The analysis of Big Data can increase the use of inactive products for collaborative economic platforms (for example, for car-sharing: predicting potential demand based on Big Data can enable optimal vehicle planning; matching algorithms and dynamic pricing algorithms based on Big Data can reduce waiting times for customers and the inactivity time of vehicles; effective re-rental strategies to rebalance the distribution of vehicles can be developed based on realistic operational data to improve the rate of rotation of the car shared between different users)
Maintenance	✓	✓	✓	✓	– Big Data can help manufacturers evaluate the general condition of a product, as well as the remaining life span of the components of the product, in order to provide an effective maintenance strategy for customers – The system based on real-time monitoring data can provide recommendations on preventing or mitigating faults – The real-time monitoring and storage of information on the product's life cycle using large volumes of data can considerably improve the degree of automation and efficacy of product maintenance

Key activities	Features of Big Data				Contribution to extending the life spans of products
Redistribution	✓	✓	✓	✓	– The system for evaluating the quality of Big Data can help companies make appropriate decisions when it comes to the reselling of returned products – An effective evaluation mechanism based on Big Data analysis can help potential buyers evaluate the reliability of second-hand sellers, thereby encouraging the development of online second-hand markets – Big Data can help second-hand platforms identify optimal marketing strategies (such as guarantees and sales prices), as well as updating and maintenance strategies, to attract more consumers and increase the sale of second-hand products – Big Data analysis can help swap and exchange platforms to permanently extend the scale of their networks of users, which can considerably increase the possibility for interactions and exchanges between strangers
Reclamation	✓	✓	✓	✓	– Simulations based on Big Data make it possible to better evaluate interventions in the life cycle of a product and can thus help develop optimal strategies, as well as assess the intervention time for updates and refurbishment – Big Data can help the manufacturer evaluate the general condition of the product and its components, which can contribute to proactive decision-making in relation to refurbishment

Table 14.3. *Features of Big Data and their contribution to extending the life spans of products*

14.3.2. *Artificial intelligence and support for decision-making in managing the life cycle of products*

Artificial intelligence (AI) integrates the production of Big Data via the Internet of Things and various data analysis methods can be applied to such data. AI makes it possible to optimize decision-making processes in conditions with multiple objectives. It offers automatic and comprehensive monitoring, thereby making significant contributions to the entire life cycle of the product, including design, manufacturing, maintenance, updating and reclamation. In particular, AI involves *intelligence*, which is to say that it does not simply follow a program but is able to reconfigure the program continuously in accordance with circumstances, using a learning and decision-making process based on previous experience. This mimicry of human thought is also known as machine learning. AI is also *automatic* in the sense that it can make decisions in an automated manner with limited, if any, human intervention. Lastly, AI is *systematic* as it does not consist of repeating a program but rather of generating new programs according to the data it acquires.

The analytical framework in Box 14.1 will help us address the possibilities that AI provides to extend the life span of products.

Improved design: a variety of Multi-Criteria Decision Aid (MCDA) systems with AI can help design products and evaluate their life cycles in a short period of time. Moreover, with the availability of faster, even quantum, and more user-friendly computers, as well as sophisticated software, it is possible to use modeling via simulations to study the life span of products effectively. In addition, data mining techniques, such as digital taxonomy, aim to categorize similar products into shared categories in order to process them in the same way with the aim of automating the evaluation of the life cycle of products. This can be done with a reasonable degree of precision and numerous constraints, even in the case of low-quality inventory data (Marwah *et al.* 2011). For example, Joshi and Gupta (2019) show that the Advanced-Remanufacturing-To-Order-Disassembly-To-Order (ARTODTO) system integrates sensors and radio-frequency identification (RFID) chips into end-of-life products in order to examine the impact of different forms of product design on reclamation. Joshi and Gupta (2019) also presented a system (ARTODTO) receiving products equipped with sensors and RFID chips in order to satisfy the demand for products, components and materials. End-of-life products can thereby be disassembled to recover the components, and can either be updated to satisfy the demand for products or recycled to satisfy the demand for materials. In particular, the system identifies designs suited to disassembly, updating and recycling in line with three criteria, namely: total profit, level of quality and the number of articles ultimately discarded by using linear physical programming to formulate the problem. For example, a new technique for modeling demand enables a company to optimize the design of its products by simultaneously considering the early stage of a product's life (manufacturing) and the later, end-of-life stage (disassembly, updating, recycling).

Maintenance and refurbishment: simulation models and techniques, such as system dynamics and discrete simulation models, can be used to evaluate the specific effects of maintenance and reclamation-related interventions on the life cycle of a product. They identify the optimal strategy and required intervention time in accordance with the actual age of a product and indicate when to update, refurbish, remake or recycle it (Charnley *et al.* 2019). The automated recommendation system is able to provide recommendations in order to prevent or mitigate faults and suggest information on the potential reasons why parts become defective (Ciora *et al.* 2015).

14.4. Conclusion

This chapter presents the main features and functions of the different technologies related to Industry 4.0. It also highlights the opportunities that these technologies present for extending the life cycles of products, whether through improved product design, better access to the product, maintenance, redistribution or reclamation. Figure 14.1 summarizes the main features of the four main technological components of Industry 4.0 that contribute to extending the life spans of products. This framework lays the groundwork for a new way of marketing products rooted in technological innovation. In particular, the IoT, Big Data and AI are strongly interconnected in reciprocal relationships, while additive manufacturing is at the heart of this process. The IoT, in fact, makes it possible to collect a greater volume of more varied and more accurate data more quickly, contributing to the development of Big Data and the related analytical capabilities. This abundance of data induced by the IoT is also feeding AI, which requires significant amounts of data to refine its algorithms and models in order to improve its decision-making processes (Delen and Ram 2018). We note that AI is also a sub-component of Big Data analysis and that it constitutes a prescriptive (what to do) analytical technique within the broader family of Big Data analysis, also including descriptive (what something is) and predictive (what will happen) data (Delen and Ram 2018). These computer technologies ultimately contribute to optimizing additive manufacturing or 3D printing, a new manufacturing paradigm relying heavily on computer support and innovation. The previous sections show how each technology contributes to extending the life span of products.

Figure 14.1 shows how they interact and how this interactivity contributes to extending the life spans. In view of the contribution of each technology to the five forms of extending the life spans of products, it seems that the IoT, Big Data, AI and additive manufacturing contribute most, through their intrinsic features, to promoting access, maintenance and redistribution; their contribution to reclamation and improved design is lower. The previous literature has already emphasized that there are comparatively fewer business models for extending the life span of products that are based on improved design than those based on maintenance, access or redistribution (Ertz *et al.* 2019). It would appear that a similar pattern is appearing in terms of the contribution of technologies to extending the life span of products. This simple statement validates *a priori*, albeit indirectly, the idea of planned obsolescence. Further means should therefore be identified to determine how Industry 4.0 could better contribute to improved design and product reclamation.

The outcomes can help managers and researchers better understand the potential for extending the life spans of products offered by technologies linked to Industry 4.0. However, we should note that the choice of a strategy for extending the life span of a product will depend on the business model of the organization in question (Ertz *et al.* 2019). If the organization provides access to products, such as car-sharing, bike-sharing or product rental companies, then the four technologies contribute to extending the life span of the product. This makes it possible to make the rented objects more profitable, increase their use rate, improve opportunities for use and thus improve the turnover. This is also the case for companies with redistribution services, such as second-hand platforms or apps, or those with resale, refurbishment, reverse logistics or remachining functions. The technologies contribute to more sustainable product design upstream, making those products easier to re-market at the end of their lives. However, these technologies also make it possible to optimize the reclamation stage by providing more information on the components of the product and its use by previous users. Many companies offering maintenance services can benefit from these technologies to optimize their operations, particularly distance maintenance carried out by users themselves or automatic maintenance, i.e. carried out by the product itself. The contributions of the technologies remain more limited for product design. However, organizations offering better-designed products, such as luxury products, would benefit from considering support from Big Data and AI and, to a lesser extent, from the IoT and additive manufacturing in order to improve the quality of their products further.

Key activities	Features and functionalities of artificial intelligence			Contributions to extending the life spans of products
	Intelligent	Automatic	Systematic	
Improvement of product design	✓	✓	✓	– An intelligent, multi-criteria support system for decision-making can help optimize product design – This makes it possible to automate the evaluation of the life cycle of a product with a reasonable precision and with relatively large constraints, which enables product design to be optimized
Access	✓	✓	✓	A highly effective and intelligent pairing system can improve the overall operational efficacy of a digital platform and the use of the product
Maintenance	✓	✓	✓	– An automated recommendation system can help prevent or mitigate faults and provide potential solutions – Through simulation, the specific effect of intervention measures in the life cycle of the product can be evaluated to determine the product strategy and the optimal intervention time
Redistribution	✓	✓	✓	– An automated recommendations system suggests to the user an element that could interest them, which could further promote the exchange of goods between individuals – Intelligent multi-actor systems can provide automated services for the analysis of semantic logistics in pairing in order to encourage transactions to be carried out in a short time period
Reclamation	✓	✓	✓	– Support a systematic decision-making framework to determine the design strategy to improve the quality of refurbished products – Automatically detect damage and propose effective strategies and measures for recovery

Table 14.4. *Features of artificial intelligence and its contribution to extending the life spans of products*

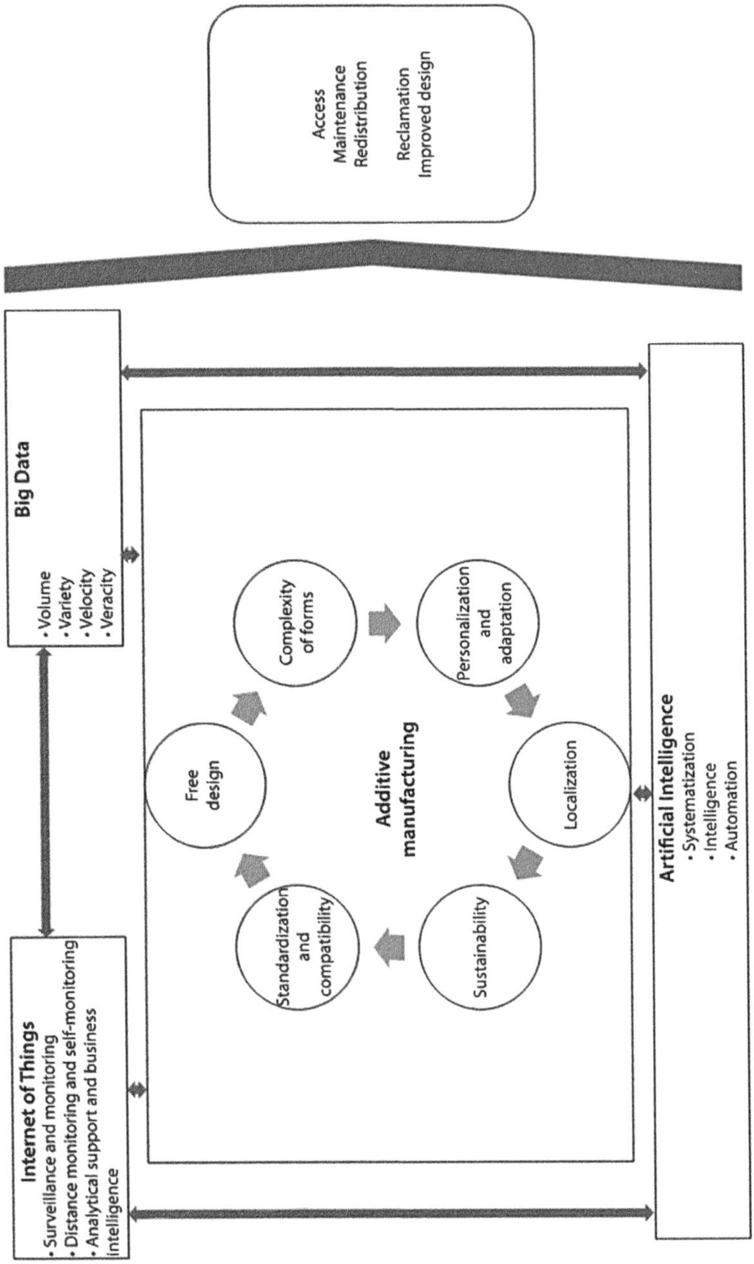

Figure 14.1. *The use of Industry 4.0 for extending the life spans of products*

More practical applications, as in the case of the company PROMPT (see Box 14.2), should be developed. Owing to differences between industries and different types of products, the scenarios of the real application of emerging technologies and their contributions can vary. It would therefore be interesting to explore these differentiating features in the future.

PROMPT is a company operating in the ICT sector, supporting the creation of partnerships, assembly and institutional financing of R&D (for example, 5G, AI, 3D simulation and virtual reality).

How do current Industry 4.0 technologies make it possible to extend the life spans of products? More specifically, how do the four major technologies in Industry 4.0 – Big Data, artificial intelligence, the Internet of Things and additive manufacturing – contribute to better designing, renting, maintaining, redistributing and recovering products?

By means of a comparative analysis of several products, we can quantify the life expectancy of products and the user experience. A good experience has the ability to reassure users of credibility through the absence of errors or the ease of use. However, the comparative analysis requires a certain level of speed in exchanging information between different inter-connected modules, hence the need to have a large and highly developed network, such as 5G. At the same time, data congestion must be reduced as far as possible, as this considerably reduces the need for calculation power and extends the life spans of machines, on the one hand, and of the end products, on the other. Moreover, the electronic management of product-related documents is aimed at better managing the life cycle of these products, which is also one of the areas of expertise of the PROMPT group. However, for proper product maintenance, the group's REACT tool makes it possible to integrate help support tools in order to monitor failures over time for each product and material. This includes, for example, the use of wireless networks, interface infrastructures and the development of new interfaces. Today, these interface platforms are in increasing demand on the market, which helps to carry out monitoring of 1,000 times as much data and to apply it to algorithms. However, this entire processing requires human support in order to maintain a certain level of monitoring and to ensure vigilance in relation to artificial intelligence. In this context, we cannot deny that there are concerns emerging among users, especially in terms of the use of personal data. The most recent example is the leak of an enormous quantity of data from the Desjardins group [a Quebecois financial cooperative]. For our part, we have developed some very promising future projects that involve the use of artificial intelligence for high levels of autonomy and performance. Some current financing requirements on the market also rely on the integration of technological innovations into products. Big organizations are also taking steps in this direction and towards this industrial revolution. For example, Ericsson is counting, on the Edge Cloud, a unique ecosystem of open and interconnected data centers, facilitating the circulation and use of data in extending the life spans of products. In short, a number of geographically distant data centers are interconnected in order to enable data to circulate in real-time at high speed. This can be used to monitor products nearby.

Box 14.2. *Interview with Luc Sirois, Director General of the PROMPT group, Quebec*

14.5. References

Charnley, F., Tiwari, D., Hutabarat, W., Moreno, M., Okorie, O., Tiwari, A. (2019). Simulation to enable a data-driven circular economy. *Sustainability*, 11(12), 3379.

Ciora, R.A., Simion, C.M., Cioca, M. (2015). Quality improvement based on Big Data analysis. *Workshop on Social Media and the Web of Linked Data*. University of Sibiu, Romania, 101–109.

Delen, D. and Ram, S. (2018). Research challenges and opportunities in business analytics. *Journal of Business Analytics*, 1(1), 2–12.

Den Hollander, M.C., Bakker, C.A., Hultink, E.J. (2017). Product design in a circular economy: Development of a typology of key concepts and terms. *Journal of Industrial Ecology*, 21(3), 517–525.

Ding, Z., Jiang, Z., Liu, Y., Wang, Y., Li, C. (2018). A Big Data based cost prediction method for remanufacturing end-of-life products. *Procedia CIRP*, 72, 1362–1367.

Ertz, M., Leblanc-Proulx, S., Sarigöllü, E., Morin, V. (2019). Made to break? A taxonomy of business models on product lifetime extension. *Journal of Cleaner Production*, 234, 867–880.

Fang, C., Liu, X., Pardalos, P.M., Pei, J. (2016). Optimization for a three-stage production system in the Internet of Things: Procurement, production and product recovery, and acquisition. *The International Journal of Advanced Manufacturing Technology*, 83(5/8), 689–710.

Informa Commercial Vehicle Staff (2018). Blockchain could help DTNA accelerate parts, maintenance [Online]. Available at: https.www.bulktransporter.com/technology/article/21657625/blockchain-could-help-dtna-accelerate-parts-main.

Jabbour, A.B.L., Jabbour, C.J.C., Foropon, C., Godinho Filho, M. (2018). When titans meet – Can industry 4.0 revolutionise the environmentally-sustainable manufacturing wave? The role of critical success factors. *Technological Forecasting and Social Change*, 132, 18–25.

Joshi, A.D. and Gupta, S.M. (2019). Evaluation of design alternatives of End-Of-Life products using Internet of Things. *International Journal of Production Economics*, 208, 281–293.

Lee, I. and Lee, K. (2015). The Internet of things (IoT): Applications, investments, and challenges for enterprises. *Business Horizons*, 58, 431–440.

Marwah, M., Shah, A., Bash, C., Patel, C., Ramakrishnan, N. (2011). Using data mining to help design sustainable products. *Computer*, 8, 103–106.

Porter, M.E. and Heppelmann, J.E. (2014). How smart, connected products are transforming competition. *Harvard Business Review*, 92, 64–89.

Research and Markets (2018). E-waste management market. Research report, July 2018 [Online]. Available at: https://www.researchandmarkets.com/research/93cbs7/opportunity?w=12 [Accessed May 25th, 2021]..

Rozières, G. (2017). Samsung, Apple et Microsoft accusés d'obsolescence programmée par Greenpeace. *Huffpost*, 27 June [Online]. Available at: https://www.huffingtonpost.fr/2017/06/27/samsung-apple-et-microsoft-accuses-dobsolescence-programmee-pa_a_23003542/.

Sauerwein, M., Doubrovski, E., Balkenende, R., Bakker, C. (2019). Exploring the potential of additive manufacturing of a product design in a circular economy. *Journal of Cleaner Production*, 226, 1138–1149.

Van Wijk, A.J.M. and Van Wijk, I. (2015). *3D Printing with Biomaterials: Towards a Sustainable and Circular Economy*. IOS Press, Amsterdam.

Conclusion

Sihem DEKHILI
CNRS – BETA, University of Strasbourg, France

This book illustrates that the values and practices of marketing and those of sustainable development can be mutually beneficial, thereby promoting new consumption patterns that are respectful of both planet Earth and collective well-being. These alternative forms of consumption are the result of new emerging expectations, and it is the role of marketing to consider consumer concerns. In the market, consumer desires and needs are naturally taken into account in developing new offers (Kotler 1971), but they are also used to question psychological ownership and alternatives to purchasing new products, forms of access-based consumption, as well as post-use behaviors.

The 14 chapters of this book perfectly show that, more than just being conceivable, the rapprochement between the fields of marketing and sustainable development is necessary. Through consideration of a multitude of sectors, examples and interviews with professionals, this book highlights how fruitful it can be to use marketing approaches and techniques to reduce barriers to responsible behaviors and encourage societal responsibility among all economic actors.

The decline of the traditional consumption pattern and the increasing difficulty of achieving profitability objectives (Moati 2018) are leading to a shift towards sustainable development and the renewed modes of exchange that sustanability requires. In this context, marketing not only provides answers in terms of product and market innovation, communications and prices, but also influences modes of exchange and distribution, the ways of participating in these exchanges and even the forms of production.

At the academic level, sustainability not only enables marketing to explore new fields of application (sustainable luxury, waste, greenbashing, ecolabels, etc.), but also helps us to reconsider existing conceptual frameworks (Extended Parallel

Process Model, Behavioral Decision Theory, etc.) and create new ones (green nudges, business models for extending the life spans of products, etc.).

While there is undeniably a link between marketing and sustainable development, it is worth highlighting and demonstrating these connections, and therein lies the relevance of this book. It provides managers, public authorities and students with proof that the spread and practices of sustainable development are indeed an issue for marketing; it also makes it possible to respond to potential reservations about a closer relationship between these two fields.

In view of what we have seen in the various contributions, here we summarize the main possibilities for associations between marketing and sustainable development, constituting promising perspectives that should be enhanced at the managerial and research levels.

Marketing is in an ideal position for encouraging change at the production and consumption levels; since its creation it has been at the heart of the relationship of exchange between the company and its customers (Lambin and Chumpitaz 2002). Thanks to its position, marketing is one of the most legitimate disciplines for questioning the compatibility of the supply and the demand. Moreover, its techniques and practices can guide individuals towards more responsible behaviors.

It therefore seems crucial that marketing studies consider ecological offers as a real alternative to conventional products; this should be at the "heart of the market" and not only a niche market intended for a narrow range of consumers. In concrete terms, priority should be given to the promotion of sustainable consumption patterns, such as packaging-free products, which require less effort and fewer sacrifices on the part of consumers. The most committed consumers can thereby act as promoters for this new market direction. As for skeptical consumers, they can be attracted by the egocentric benefits of the ecological offer, in addition to its altruistic value. In this sense, marketing has a major role to play, as its purpose has always been to create value for the stakeholders involved in the exchange relationship and to enhance it among the target customers (American Marketing Association 2007). Marketing thus offers a benefit by strengthening the commercial appeal of responsible products and expanding the consumer segment beyond individuals who already express strong environmental and social concerns. By using levers such as price and communication, marketing is able to project a certain level of quality and value onto a given item, which has an impact on the way the consumer will consider the product or service throughout its lifecycle, the use that will be made of it, the risk of waste associated with it and so on.

The product (and/or service) is the focus of marketing and is, at the same time, the subject that has fuelled debates around sustainable development: what is the product's environmental and social cost? Does it respond to a consumer need?

In terms of mindful consumption, or even the trend towards deconsumption, marketing is able to act on the characteristics of the offer and its impact on the planet. In fact, enhancing the value of the ecological materials used is one of the challenges that marketing is quite capable of overcoming, and can be facilitated by the use of innovation. The technological revolution may help to offer augmented sustainable products and facilitate the emergence of new ecological materials (such as vegetable leather to replace animal leather) that are more easily promoted to consumers. Moreover, the rise of digital applications offers various possibilities to marketing, as they can be used to engage people and encourage them to adopt ecological practices, such as recycling, selecting responsible materials and brands and combating waste. However, faced with new products and consumption patterns, consumers are expressing a need for information and guidance. The result of this is that marketing is reconsidering the role of the salesperson; while the scope of their tasks has been reduced in recent years, the roles of sellers are tending to become more varied in light of the current trend for sustainability (advisor, supporter, guarantor of quality, etc.). This permits salespersons not only to guide consumers in their decision-making processes, but also to preserve the properties of the ecological products offered.

Owing to its interest in the study of consumer behavior, marketing provides a thorough understanding of the sources of valorization of responsible consumption, as well as of the reasons for consumer skepticism and resistance to changing their behaviors. By using a variety of approaches and frameworks, it helps us to analyze the most significant barriers, including psychological ones. In addition, in the context of the trend for sustainability, marketing is able to define what falls within the consumer's responsibility and what is a matter for other actors.

Through the opportunities for segmentation it offers, marketing makes it possible to take into account the heterogeneity of individuals and their different levels of ecological awareness, which has the advantage of increasing the effectiveness of responsible offers and the acceptance of public measures in favor of sustainable development. Thanks to its operational capacity, marketing provides a range of appropriate actions that could influence the values, decisions and choices of each consumer segment.

Another key area in marketing is communication. Identifying the themes that should be communicated, as well as the tone that should be used to spread the message to the target audience, constitute fundamental issues in marketing. Communication based on the values of sustainable development thus becomes a vector for an improved brand image. Marketing has the advantage of identifiying the credibility and effectiveness determinants of a communication, offering a significant added value to consumers. It is also able to ensure effective communication strategies about complex topics, such as climate change. In this regard, companies should adopt clear and sincere initiatives to avoid any risk of confusion or accusations of opportunism. For example, despite LVMH having been the first brand to adapt its production line to produce hydroalcoholic gel for health care personnel during the COVID-19 crisis, it was the subject of severe criticism on social media owing to its logo having been put on the product's packaging.

In a context of increasing consumer mistrust of corporate practices, companies are placing the priority on protecting their brand equity and building and/or consolidating links to environmental and social responsibility. Sustainable development is becoming part of a brand's heritage (Vanhamme and Grobben 2009).

Nowadays, consumers are expressing a need to give meaning to their consumption; they are increasingly sensitive to brand values and the convictions of business leaders, who they expect to make significant and tangible contributions to sustainability. Monetary donations to a cause, which is a common practice among companies looking to "green" their image, often appears to be the easy option. In this sense, marketers should pay particular attention to the link between the cause championed by companies and their sectors of activity. For example, the "courage is beautiful" advertising campaign by Dove Canada during the COVID-19 crisis, which used photos of hospital workers whose faces showed marks made by protective masks, reaffirmed the brand's long-standing campaign for "real beauty".

In order to reassure consumers of the sincerity of their commitment, many companies have used control mechanisms. By promoting ecolabels, considered to be a selling point, marketers can contribute to strengthening the credibility of responsible offers. As with consumers, ecolabeling procedures allow producers, including those in countries with low commitments to sustainability (as in West African countries), to embrace ecological practices. It is also important to take into account the correspondence between the value of the offer and the proposed price; the price policies should be perceived as fair by consumers and not associated with opportunistic behavior by companies seeking to take advantage of market demand. In this way, marketing can continue to be effective while also being ethical and responsible!

Lastly, the potential for possible associations between marketing and sustainable development is far from being exhausted. The digital revolution in particular offers innumerable opportunities that could lead to reconsideration of the ways in which ecological offers are produced, marketed and distributed, as well as the ways of influencing consumer decisions. Ongoing digital developments are moving towards increasingly personalized marketing, which also encourages strong digital interactions between actors. Taken together, this can lead to a stronger commitment to sustainable development among individuals. Beyond the personalization of mobile apps (ergonomics, notifications, etc.), technology 4.0 is able to adapt information about ecolabels to suit consumer profiles. In distribution, deep learning can help the consumer, through intelligent theatricalization, to benefit from the improved accessibility of ecological products in stores.

That being said, the major challenge in the coming years will still be convincing people of the good intentions of marketers in regard to sustainable development. Even there, technology is of considerable interest: blockchains can ensure the accuracy of information about ecological products and thus increase consumer trust.

This means that the strengthening of links between marketing and sustainable development cannot ignore the societal and technological transformations that characterize our present! It is without doubt these changes that will facilitate a broader dissemination of responsible marketing in our societies.

References

American Marketing Association (2007). Definition of marketing [Online]. Available at: https://www.ama.org/listings/2007/12/24/definition-of-marketing 2/.

Kotler, P. (1971). Social marketing: An approach to planned social change. *Journal of Marketing*, 35(3), 3–12.

Lambin, J.J. and Chumpitaz, R. (2002). Marketing stratégique et opérationnel : du marketing à l'orientation-marché. Dunod, Paris.

Moati, P.H. (2018). (Dé)penser la consommation : peut-il y avoir une "bonne consommation" ? EMS, Caen.

Vanhamme, J. and Grobben, B. (2009). "Too good to be true", The effectiveness of CSR history in "countering negative publicity". *Journal of Business Ethics*, 85(S2), 273–283.

List of Authors

Mohamed Akli ACHABOU
IPAG Business School
Paris
France

Abdelmajid AMINE
IRG
University Paris-Est Créteil
France

Mouna BENHALLAM
IRG
University Paris-Est Créteil
France

Laurent BERTRANDIAS
Toulouse Business School
France

Émilie BOILY
LaboNFC
University of Quebec at Chicoutimi
Saguenay
Canada

Gaëlle BOULBRY
LP3C (Laboratoire de psychologie, cognition, comportement, communication)
University of South Brittany
Vannes
France

Julien BOUSQUET
LaboNFC
University of Quebec at Chicoutimi
Saguenay
Canada

Géraldine CHABOUD
MOISA – Cirad
Montpellier
France

Amélie CLAUZEL
Paris 1 Panthéon-Sorbonne University
ISC Paris
France

Mantiaba COULIBALY-BALLET
GRM
Côte d'Azur University
IAE Nice
France

Maud DANIEL-CHEVER
CREM
CNRS
IGR-IAE Rennes
University of Rennes 1
France

Sihem DEKHILI
CNRS – BETA
University of Strasbourg
France

Mickaël DUPRÉ
Governance and Insular Development
University of French Polynesia
France

Margot DYEN
LEGO (Laboratoire d'économie et de gestion de l'Ouest)
University of Western Brittany
Quimper
France

Leila ELGAAIED-GAMBIER
Toulouse Business School
France

Samer ELHAJJAR
University of Balamand
Tripoli
Lebanon

Myriam ERTZ
LaboNFC
University of Quebec at Chicoutimi
Saguenay
Canada

Jean-Marc FERRANDI
LEMNA
ONIRIS
Nantes University
France

Sylvie FOUTREL
LEGO (Laboratoire d'économie et de gestion de l'Ouest)
Catholic University of West Brittany South
Arradon
France

Agnès FRANÇOIS-LECOMPTE
LEGO (Laboratoire d'économie et de gestion de l'Ouest)
University of South Brittany
Vannes
France

Patrick GABRIEL
LEGO (Laboratoire d'économie et de gestion de l'Ouest)
University of Western Brittany
Brest
France

Christine GONZALEZ
Argumans
University of Le Mans
France

Nathalie GUICHARD
RITM
Paris-Saclay University
Sceaux
France

Damien HALLEGATTE
LaboNFC
University of Quebec at Chicoutimi
Saguenay
Canada

Yassine LAGHRIB
LaboNFC
University of Quebec at Chicoutimi
Saguenay
Canada

Imen LATROUS
LaboNFC
Université du Québec à Chicoutimi
Saguenay
Canada

Guillaume LE BORGNE
IREGE
University of Savoie Mont Blanc
Annecy
France

Élisa MONNOT
THEMA
CNRS
CY Cergy Paris University
France

Manu NAVARRO
REGARDS
University of Reims Champagne-Ardenne
France

Adeline OCHS
Audencia Business School
Nantes
France

Philippe ODOU
REGARDS
University of Reims Champagne-Ardenne
France

Kubiat PATRICK
LaboNFC
University of Quebec at Chicoutimi
Saguenay
Canada

Gautier Georges Yao QUENUM
LaboNFC
University of Quebec at Chicoutimi
Saguenay
Canada

Fanny RENIOU
CREM-CNRS
IGR-IAE Rennes
University of Rennes 1
France

Caroline RICHÉ
RITM
Paris-Saclay University
Sceaux
France

Marie SCHILL
REGARDS
University of Reims Champagne-Ardenne
France

Julien SCHMITT
ESCP Business School
Paris
France

Maxime SEBBANE
Research Center
Paul Bocuse Institute
Écully
France

Béatrice SIADOU-MARTIN
MRM
University of Montpellier
France

Lucie SIRIEIX
UMR MOISA – CIRAD
University of Montpellier
CIHEAM-IAMM
INRAE
Institut Agro-IRD
Montpellier
France

Shouheng SUN
LaboNFC
University of Quebec at Chicoutimi
Saguenay
Canada

John THØGERSEN
Department of Management
Aarhus University
Denmark

Index

C

G, H

I

L

M

Q, R

S